ENVIRONMENTAL REMEDIATION TECHNOLOGIES, REGULATIONS AND SAFETY

BALLAST WATER AND INVASIVE SPECIES

MANAGEMENT AND TREATMENT EFFICACIES

ENVIRONMENTAL REMEDIATION TECHNOLOGIES, REGULATIONS AND SAFETY

Additional books in this series can be found on Nova's website under the Series tab.

Additional E-books in this series can be found on Nova's website under the E-book tab.

ENVIRONMENTAL REMEDIATION TECHNOLOGIES, REGULATIONS AND SAFETY

BALLAST WATER AND INVASIVE SPECIES

MANAGEMENT AND TREATMENT EFFICACIES

GILLES WUYTS
EDITOR

nova publishers
New York

For permission to use material from this book please contact us:
Telephone 631-231-7269; Fax 631-231-8175
Web Site: http://www.novapublishers.com

Additional color graphics may be available in the e-book version of this book.

Library of Congress Cataloging-in-Publication Data

ISBN: 978-1-62618-754-2

Published by Nova Science Publishers, Inc. † New York

CONTENTS

PREFACE

With increases in the number of people traveling, the speed and methods of travel, the types and volume of trade, the ability to move living plants and animals so that more of them survive the journey, and the different modes of transport for hitch-hiking organisms, invasive species have become a global concern. Although there are many ways in which species may invade, this book focuses on ballast water discharge by cargo ships as one of the more significant mechanisms for biotic invasion of coastal and estuarine habitats as well as inland navigable waters. The economic, social, recreational, and ecological losses/costs attributable to aquatic invasive species are difficult to quantify. While some costs have been estimated, such as the $5 billion in damages to water pipes, boat hulls, and other hard surfaces by zebra mussels in the Great Lakes, others, such as the losses of native species and environment restoration to pre-invasion quality, are unknown.

Chapter 1 – In recent years, people have become increasingly aware that the globalization of trade, the increased speed of travel, the massive volume of cargo shipments, and rising tourism have combined to increase the chance of accidental introductions of foreign species into the United States. Aquatic species arrive through a variety of mechanisms—unintentionally when attached to vessel hulls or carried in vessel ballast water and intentionally when imported for aquaria display, as live seafood for human consumption, or as a transplant to increase sport fishing opportunities.

The arrival of zebra mussels in the Great Lakes and their subsequent damage to city water supplies and electric utilities focused significant attention on ballast water discharge by cargo ships as a high-risk mechanism for species invasion. New management efforts attempt to address this concern.

On March 23, 2012, the U.S. Coast Guard published a final rule establishing quantitative standards for ballast water treatment. The new standards initially follow standards developed by the International Maritime Organization (IMO). In a subsequent phase, the quantitative standards would become much more stringent, given sufficient technological development to support achievement of the higher standards. The Coast Guard standards do not preempt existing state ballast water management standards.

In response to litigation, the Environmental Protection Agency (EPA) published regulations on December 28, 2008, to regulate ballast water discharge under the Clean Water Act through vessel general permits (VGPs). Environmental groups expressed concern over the effectiveness of the EPA regulations and filed a court challenge. In a March 2011 settlement, EPA was required to revise its VGP program to require numeric limits on organisms in ballast water discharges as well as additional monitoring and reporting of vessel discharges. Under the terms of this settlement, EPA proposed two new VGP programs on November 30, 2011. Meanwhile, P.L. 110-299 provided a two-year moratorium from the VGP program for commercial fishing vessels and non-recreational vessels less than 79 feet in length, and the 111th Congress enacted legislation (P.L. 111-215) to further extend this moratorium through December 18, 2013. EPA expects to take final action on the draft permits by November 30, 2012.

In addition to federal programs and sometimes in response to perceived deficiencies in federal regulation, several states have chosen to regulate aspects of ballast water management. Concerns have arisen over particularly stringent state ballast water regulations as to whether the technology exists to meet their standards. In the 112th Congress, Section 459 of H.R. 2584, as reported by the House Committee on Appropriations, would prohibit EPA from providing funds to any Great Lakes state that has a more stringent performance or ballast water exchange standard than either a revised Coast Guard standard or the IMO standard.

This report provides background on various approaches to ballast water management and reviews current ballast water management laws and programs.

Chapter 2 – Vessel ballast water discharges are a primary source of introductions of nonindigenous species and potentially harmful pathogens to marine, estuarine, and freshwater ecosystems of the United States. At present, federal requirements for managing ballast water discharges rely primarily on ballast water exchange. However, the U.S. Coast Guard (USCG) has proposed numeric concentration-based limits for live organisms in ballast water that would initially require compliance with a "Phase 1 standard" that has the same concentration limits as the International Maritime Organization (IMO) D-2 standard and subsequently require compliance with a more stringent "Phase 2 standard." In addition, the U.S. Environmental Protection Agency is considering numeric standards that limit the number of live organisms in discharged ballast water in its revision of the Vessel General Permit.

This Advisory report responds to a request from the EPA Office of Water (OW) for EPA's Science Advisory Board (SAB) to provide advice on technologies and systems to minimize the impacts of invasive species in vessel ballast water discharge. More specifically, the SAB was requested to provide review and advice regarding whether existing shipboard treatment technologies can reach specified concentrations of organisms in vessel ballast water, how these technologies might be improved in the future, and how to overcome limitations in existing data. To conduct the assessment, the SAB's Ecological Processes and Effects Committee (EPEC) was augmented with additional Panel members (Ballast Water Advisory Panel, or "Panel") having expertise in ballast water issues, marine engineering, and engineering treatment technologies.

In addition to the SAB assessment of ballast water management systems (BWMS), the EPA and the USCG commissioned the National Academy of Sciences (NAS) to conduct a complementary study that assesses the risk of successful invasions as a function of different concentrations of organisms in ballast water discharges. Therefore, this SAB Advisory does not address the relationship between the various concentrations of organisms as described in proposed standards and the likelihood of invasions or the potential effects of pathogens. Rather, in this report, "effectiveness" refers to how well a technology may meet a given numeric concentration limit for organisms, not to how well the technology may protect the environment. Similarly, phrases such as "meets more stringent" or "meets less stringent" proposed standards are used as descriptive conclusions and do not represent performance recommendations.

To prepare this Advisory report, the Panel reviewed a "Background and Issue Paper" written by EPA's OW and USCG (Albert et al. 2010). This paper provided an overview of information about major categories of shipboard ballast water treatment technologies and presented proposed ballast water discharge standards drawn from international sources, the USCG, and nine states. In addition, EPA's OW and the public identified information on 51 existing or developmental ballast water treatment technologies, although detailed data were available for only 15 specific BWMS. The Panel used this information as the source material for its assessment of ballast water treatment performance and, as requested by the EPA, used proposed ballast water discharge standards as the performance benchmarks.

In: Ballast Water and Invasive Species
Editor: Gilles Wuyts

ISBN: 978-1-62618-754-2

Chapter 1

BALLAST WATER MANAGEMENT TO COMBAT INVASIVE SPECIES*

Eugene H. Buck

SUMMARY

In recent years, people have become increasingly aware that the globalization of trade, the increased speed of travel, the massive volume of cargo shipments, and rising tourism have combined to increase the chance of accidental introductions of foreign species into the United States. Aquatic species arrive through a variety of mechanisms—unintentionally when attached to vessel hulls or carried in vessel ballast water and intentionally when imported for aquaria display, as live seafood for human consumption, or as a transplant to increase sport fishing opportunities.

The arrival of zebra mussels in the Great Lakes and their subsequent damage to city water supplies and electric utilities focused significant attention on ballast water discharge by cargo ships as a high-risk mechanism for species invasion. New management efforts attempt to address this concern.

On March 23, 2012, the U.S. Coast Guard published a final rule establishing quantitative standards for ballast water treatment. The new standards initially follow standards developed by the International Maritime Organization (IMO). In a subsequent phase, the quantitative standards would become much more stringent, given sufficient technological development to support achievement of the higher standards. The Coast Guard standards do not preempt existing state ballast water management standards.

In response to litigation, the Environmental Protection Agency (EPA) published regulations on December 28, 2008, to regulate ballast water discharge under the Clean Water Act through vessel general permits (VGPs). Environmental groups expressed concern over the effectiveness of the EPA regulations and filed a court challenge. In a March 2011 settlement, EPA was required to revise its VGP program to require numeric limits on organisms in ballast water discharges as well as additional monitoring and reporting of vessel discharges. Under the terms of this settlement, EPA proposed two new VGP programs on November 30, 2011. Meanwhile, P.L. 110-299 provided a two-year

* This is an edited, reformatted and augmented version of a Congressional Research Service publication, CRS Report *for Congress* RL32344, *from* www.crs.gov, *p*repared for Members and Committees of Congress, *dated* April 10, 2012.

moratorium from the VGP program for commercial fishing vessels and non-recreational vessels less than 79 feet in length, and the 111th Congress enacted legislation (P.L. 111-215) to further extend this moratorium through December 18, 2013. EPA expects to take final action on the draft permits by November 30, 2012.

In addition to federal programs and sometimes in response to perceived deficiencies in federal regulation, several states have chosen to regulate aspects of ballast water management. Concerns have arisen over particularly stringent state ballast water regulations as to whether the technology exists to meet their standards. In the 112th Congress, Section 459 of H.R. 2584, as reported by the House Committee on Appropriations, would prohibit EPA from providing funds to any Great Lakes state that has a more stringent performance or ballast water exchange standard than either a revised Coast Guard standard or the IMO standard.

This report provides background on various approaches to ballast water management and reviews current ballast water management laws and programs.

INTRODUCTION

With increases in the number of people traveling, the speed and methods of travel, the types and volume of trade, the ability to move living plants and animals so that more of them survive the journey, and the different modes of transport for hitch-hiking organisms, invasive species have become a global concern. Although there are many ways in which species may invade,[1] this report focuses on ballast water discharge by cargo ships as one of the more significant mechanisms for biotic invasion of coastal and estuarine habitats as well as inland navigable waters.

The arrival of zebra mussels in the Great Lakes in the late 1980s and their subsequent damage to city water supplies and electric utilities[2] focused initial attention on ballast water as a source of invasive species. Reflecting the scope of the problem, the San Francisco Bay-Delta ecosystem is considered to be one of the most disrupted aquatic ecosystems in the United States, with colonization by more than 230 non-native species.[3] Compounding the problem, species that invaded via other mechanisms (e.g., mitten crabs that are believed to have been an illegal seafood introduction) may be further spread through ballast water transport and vice versa (e.g., zebra mussels are spreading to new drainages via recreational boats transported by highway on trailers). In the Gulf of Mexico, ballast water has been implicated in the contamination of commercial oyster beds.[4] Globally, it is estimated that more than 10,000 marine species each day may be transported across the oceans in the ballast water of cargo ships.[5] The ctenophore *Mnemiopsis leidyi* was transported in ballast water from estuaries along the Atlantic coast of North America to the Black Sea, where it is blamed for a massive collapse of the commercial fish harvest. Around 1978, the American jack-knife clam, *Ensis directus*, was introduced through ballast water in the German Bight and has spread rapidly over the North Sea coast, where it has become one of the most common bivalves, replacing many native species.

The economic, social, recreational, and ecological losses/costs attributable to aquatic invasive species are difficult to quantify. While some costs have been estimated, such as the $5 billion in damages to water pipes, boat hulls, and other hard surfaces by zebra mussels in the Great Lakes,[6] others, such as the losses of native species and environment restoration to pre-invasion quality, are unknown. The 112th Congress may consider legislative proposals to reauthorize and amend the Nonindigenous Aquatic Nuisance Prevention and Control Act of 1990,[7] including specific provisions that would modify how ballast water is managed.

BALLAST WATER MANAGEMENT

Ballast water is water that is held in tanks and/or cargo holds of ships to provide stability and maneuverability during a voyage when ships are not carrying cargo, are not carrying heavy enough cargo, or require more stability due to rough seas.[8] Ballast water may be either fresh or saline. Ballast water may also be carried so that a ship rides low enough in the water to pass under bridges and other structures. Ballast water management (BWM) for vessels includes all measures that aim to prevent unwanted aquatic nuisance species from being transported between ports in the ballast. Seaports in which ships exchange ballast water daily are at severe risk of invasions. Organisms transported to U.S. ports from foreign harbors with similar physiochemical characteristics (e.g., water temperatures, salinity regimes) pose an especially high risk of invasion. Even if only a tiny proportion of newly arriving non-native species survive in new habitats, such as San Francisco Bay, Chesapeake Bay, or Boston Harbor, the actual number of successful invasive species can be very large.

There are several different ways of managing ballast water. Currently, the most widely used is ballast water exchange. Ballast water exchange means that ships on their way to the next port release the lower-salinity coastal water they brought aboard in their last port and replace it with higher-salinity open-ocean water. Although this measure is not perfect, it reduces the number of potentially invasive species in the ballast tanks and replaces them with oceanic organisms that are less likely to survive in the lower-salinity near-shore waters of the ship's next port.[9] However, organisms with a wide tolerance for differing salinities may survive ballast water exchange, especially any such organisms that may reside in the unpumpable residual water and sediment remaining in the tanks during any ballast water exchange.

Another approach to BWM is treatment.[10] Ballast water treatment is the subject of extensive current research and development, and several technologies and methodologies have been proposed. These include mechanical methods (e.g., filtration and separation), physical methods (e.g., sterilization by ultraviolet light, ozone, heat, electric current, or ultrasound), and chemical methods (using biocides). In addition, treatment may combine several of these methods.[11]

Treatment may be an appropriate management option on occasions when vessels temporarily operate without ballast—a "no-ballast-on-board" (NOBOB) situation. When a ship is carrying no ballast water, it presents unique treatment problems because large numbers of organisms can reside in the unpumpable residual water and sediment remaining in the ballast tanks. Few of the tested methodologies have been applied to the control of organisms in NOBOB situations.[12] The treatment option favored by many ship operators because of its intrinsic simplicity and relatively low cost is biocide application, whereby chemical agents are added to the ballast water to minimize the number of (i.e., kill) viable organisms. This approach also has the potential to address the NOBOB condition.[13] Concerns remain relating to establishing and enforcing standards for the appropriate disposal of biocide-treated ballast water and sediments.

Although estimates of the costs of ballast treatment may be imprecise and vary from vessel to vessel, there is some general agreement on average costs.[14] For example, it may cost an estimated $400,000 per vessel for modification of container/bulk vessels to use onshore ballast water treatment facilities at California ports. More generally, the cost of retrofitting vessels to treat ballast water has been estimated at between $200,000 and $310,000 per vessel for mechanical treatment and around $300,000 for chemical treatment.[15] Most of this expense will be borne by foreign shipping companies, as the U.S. flag fleet is a small percentage of the global fleet,[16] and likely passed along to consumers of products imported on these ships. The likelihood of compliance by the foreign flag fleet was increased by the February 2004 conclusion of an international agreement on ballast water management (see "International Efforts," below).

CURRENT U.S. LAW

Attempts to address ballast water concerns in the United States began with the Nonindigenous Aquatic Nuisance Prevention and Control Act of 1990 (NANPCA),[17] which established a federal program to prevent the introduction and to control the spread of unintentionally introduced aquatic nuisance species. The U.S. Coast Guard, U.S. Environmental Protection Agency (EPA), U.S. Fish and Wildlife Service (FWS), Army Corps of Engineers, and National Oceanic and Atmospheric Administration (NOAA) share responsibilities for implementing this effort, acting cooperatively as members of an Aquatic Nuisance Species (ANS) Task Force to conduct studies and report to Congress to (1) identify areas where ballast water exchange can take place without causing environmental damage; and (2) determine the need for controls on vessels entering U.S. waters other than the Great Lakes. Under §1101 of NANPCA, a Great Lakes BWM program (voluntary in its first two years) became mandatory in 1992. This section directed the Coast Guard to issue regulations (33 CFR Part 151, Subpart C) to prevent the introduction and spread of aquatic nuisance species into the Great Lakes through the ballast water of vessels and established civil and criminal penalties for violating these regulations. NANPCA also encouraged the Secretary of Transportation (now the Secretary of Homeland Security) to negotiate with foreign countries, through the International Maritime Organization, to prevent and control the unintentional introduction of aquatic nuisance species.

In 1996, the National Invasive Species Act (NISA) amended NANPCA to create a national ballast management program modeled after the Great Lakes program wherein all ships entering U.S. waters (after operating outside the U.S. Exclusive Economic Zone) are directed to undertake high seas (i.e., mid-ocean) ballast exchange or alternative measures pre-approved by the Coast Guard as equally or more effective. While not initially enforced on a ship-by-ship basis, this national program was to have become mandatory within three years of the date the Coast Guard issued interim voluntary guidelines[18] if ships did not show adequate compliance with the program[19] in the absence of enforcement.

The National Ballast Information Clearinghouse (NBIC) was developed jointly by the Coast Guard and the Smithsonian Environmental Research Center to synthesize, analyze, and interpret national data concerning BWM.[20] During the first two years (July 1999 through June 2001), the NBIC found that nationwide compliance with ballast exchange reporting requirements was low, with only 30.4% of vessels entering the U.S. Exclusive Economic Zone (EEZ) filing reports with the NBIC.[21] In addition, a Coast Guard Report to Congress concluded that reporting compliance was insufficient to allow an accurate assessment of voluntary BWM.[22] On January 6, 2003, the Coast Guard proposed penalties for those who failed to submit BWM reports required by 33 U.S.C. §151 Subpart D for most vessels entering U.S. waters.[23] On June 14, 2004, the Coast Guard published final regulations establishing penalties for failure to comply with reporting requirements.[24] On July 28, 2004, the Coast Guard published final regulations changing the voluntary ballast water management program to a mandatory one.[25]

NISA encouraged negotiations with foreign governments to develop and implement an international program for preventing the introduction and spread of invasive species in ballast water. NISA also required a Coast Guard study and report to the Congress on the effectiveness of existing shoreside ballast water facilities used by crude oil tankers in the coastal trade off Alaska, as well as studies of Lake Champlain, the Chesapeake Bay, San Francisco Bay, Honolulu Harbor, the Columbia River system, and estuaries of national significance.[26] Under NISA, a Ballast Water Management Demonstration Program was authorized to promote the research and development of technological alternatives to ballast water exchange.

NISA has been criticized as inadequate and faulted for several alleged shortcomings, including agency weakness or delay in implementing some of its provisions.[27] Since NISA exempted most coastwise vessel traffic from ballast water exchange guidelines, vessels traveling short distances between U.S. ports (e.g., from San Francisco Bay, which is highly invaded, to Puget Sound, which is less so) are exempt from controls. Others are critical of the provisions of 16 U.S.C. §4711(k)(2)(A) giving the vessel owner a blanket exemption to ignore any mandatory regulations if the master determines that the vessel might not be able to safely conduct a ballast water exchange on the open ocean. Whereas earlier provisions applicable to the Great Lakes provided a safety exemption, the master/captain of a vessel was required to report the problem to the Coast Guard and conduct alternate BWM measures, often negotiated on a case-by-case basis. Critics believe the NISA language has eliminated any incentive to change ballast water piping systems or adopt other management or treatment options to deal with the problem safely. Finally, NISA has been criticized for its apparent failure to actually prevent additional introductions of damaging organisms into the Great Lakes, despite this being the one area where the requirements for managing ballast water have been the most stringent for the longest time.[28]

Federal Agency Programs

U.S. Coast Guard

Under NANPCA, the Coast Guard is responsible for developing and implementing a BWM program to prevent the unintentional introduction and dispersal of nonindigenous aquatic species into waters of the United States from ship ballast water.[29] Initially this was accomplished through a mandatory BWM program for the Great Lakes ecosystem and voluntary guidelines for the remainder of U.S. waters. Relevant regulations, published at 33 CFR Part 151, Subpart C, went into effect in 1993.[30]

Under these regulations, the Coast Guard enforced mandatory requirements for ballast water management only for the Great Lakes. Ballast water reporting data for inbound vessels was submitted via fax either directly to the Captain of the Port (COTP) Buffalo or U.S. Coast Guard's Marine Safety Detachment (MSD) Massena, or via the Saint Lawrence Seaway Development Corporation (SLSDC) to MSD Massena at least 24 hours before arrival (33 CFR 151.2040). Compliance with these requirements was essentially 100%. Every vessel that reports that it is carrying ballast water on board (BOB) while transiting to the Great Lakes underwent ballast water inspections by either MSD Massena at the locks in Massena, NY, or by the SLSDC in Montreal. Compliance with the mandatory ballast water requirements averaged approximately 92% in the late 1990s and early 2000s for those vessels declaring BOB. Those vessels found to not have conducted proper exchange were ordered not to discharge ballast water in the Lakes. For these vessels, or if the vessel declared that it intended to retain its ballast water on board while in the Great Lakes system, then MSD Massena again boarded the vessel after it has made port calls in the Lakes and passed the lock in Massena on its outbound transit to ensure the vessel had not discharged while in the Lakes. Vessel compliance with the order not to discharge has been excellent.[31]

As stated in the Secretary of Transportation's November 2001 *United States Coast Guard Report to Congress on the Voluntary National Guidelines for Ballast Water Management*, the Coast Guard began developing regulations requiring active BWM of all ships that enter U.S. waters after operating beyond the EEZ, including sanctions for failure to comply. On July 30, 2003, the Coast Guard proposed mandatory BWM practices for all vessels with ballast tanks bound for ports or places within the United States or entering U.S. waters, but excluding domestic port-toport voyages.[32] The Coast Guard published final regulations on July 28, 2004, extending BWM requirements already in place for vessels entering the Great Lakes and Hudson River to all U.S. ports and waters.[33] The Coast Guard requested comments on potential revisions to mandatory ballast water management reporting requirements in early 2007.[34]

The Coast Guard also has taken action to establish a quantitative ballast water treatment performance standard; protocols for testing, verifying, and reporting on treatment technologies; and a program to facilitate experimental shipboard installation and operation of promising technologies. On March 4, 2002, the Coast Guard published an advance notice of proposed rulemaking, seeking comments on development of a ballast water treatment goal and an interim ballast water treatment standard as part of regulations that would make guidelines for ballast exchange mandatory.[35] On September 26, 2003, the U.S. Coast Guard announced its intent to prepare a Programmatic Environmental Impact Statement for proposed regulatory action to establish a ballast water discharge standard.[36] Such a standard would establish the required level of environmental protection necessary to prevent introductions and combat the spread of invasive species from ballast water discharges. Comments were accepted on this proposal through December 26, 2003. On May 2, 2006, the U.S. Coast Guard published a request for public comment on the status of research and development of ballast water management systems and analytical technologies for testing such systems.[37] On August 28, 2009, the Coast Guard published proposed regulations to establish standards for ballast water treatment.[38] On March 23, 2012, the Coast Guard published its final rule.[39] The new standards initially follow standards developed by the

International Maritime Organization (see "International Efforts," below), and would become much more stringent in a subsequent phase, given sufficient technological development to support compliance with higher standards.[40] The new ballast water discharge standard is effective 90 days after publication, or June 21, 2012. The Coast Guard standards do not preempt state ballast water management standards (see "State Programs," below).

On January 2, 2004, the U.S. Coast Guard announced the beginning of a program to facilitate the installation of experimental shipboard ballast water treatment systems on both foreign and domestic vessels.[41] This Shipboard Technology Evaluation Program (STEP) aims to promote research and development of shipboard ballast water treatment systems through regulatory incentives, creating more options for vessel owners seeking alternatives to ballast water exchange. Regulatory incentives would grant conditional equivalencies for accepted vessels in the STEP that might not meet discharge standards mandated by future regulations. On August 5, 2004, the U.S. Coast Guard announced its interest in establishing a program to approve ballast water treatment systems to assure that approved systems meet ballast water discharge standards the Coast Guard anticipates implementing.[42] On August 28, 2009, the Coast Guard published proposed regulations that would establish an approval process for ballast water management systems.[43] In September 2010, the Coast Guard, EPA, and U.S. Navy published *Generic Protocol for the Verification of Ballast Water Treatment Technology*.[44]

On August 31, 2005, the Coast Guard published a notice of policy establishing best management practices for NOBOB vessels entering the Great Lakes that have residual ballast water and ballast tank sediment.[45]

Department of State

The Department of State participated in negotiations through the International Maritime Organization to develop an international convention to control the spread of invasive species from the exchange of ships' ballast water. (See "International Efforts," below.)

National Oceanic and Atmospheric Administration

Through both the National Sea Grant College program and a BWM technology development program, NOAA has funded research on alternatives to ballast water exchange as methods of BWM.[46] NOAA's Great Lakes Environmental Research Lab (GLERL) targets prevention and control to stop the inflow and spread of new aquatic organisms, with particular emphasis on ship ballast. GLERL, with combined funding from NOAA and several other agencies, developed and provides leadership for the Great Lakes NOBOB and Ballast Exchange research program, focusing on the biological assessment of ballast tank residuals and the experimental determination of effectiveness of ballast exchange.

In this program, GLERL scientists collaborate with scientists at several universities and the Smithsonian Environmental Research Center. In a related project, scientists at GLERL and the University of Michigan are evaluating two chemicals for use on residuals in NOBOB tanks. In addition, GLERL has been working on developing a model of ballast tank flow during ballast tank exchange.

NOAA also has managed a Ballast Water Management Demonstration Program grant competition,[47] but no recent grant announcements have been published.

Department of Defense

The Department of Defense (DOD) is promulgating joint regulations with the EPA covering discharges from DOD vessels (40 CFR 1700) to implement §312(n) of the Clean Water Act (33 U.S.C. §1322(n)). When complete, they will set discharge standards for vessel ballast water to address the environmental effect of non-native species introduction via that ballast water (as well as addressing chemical pollution from other Armed Forces vessel discharges).

The regulations are being developed in three phases. The first, completed in May 1999, determined which ballast water discharges would require control.

The second, currently in progress, will set performance standards, and the third will promulgate regulations for meeting those standards.

Smithsonian Institution

The Smithsonian Environmental Research Center performs research to examine patterns of ballast water delivery and measures species transfer associated with shipping. In cooperation with the Coast Guard, the center established the National Ballast Information Clearinghouse (NBIC)[48] to measure the changing patterns of ballast water delivery and management for vessels arriving in U.S. ports and to synthesize national data on patterns and impacts of alien species in coastal ecosystems.

Environmental Protection Agency

The EPA establishes water quality standards and regulates discharges under the Clean Water Act.[49] On September 9, 2003, the EPA announced that it was denying a 1999 petition by the Pacific Environmental Advocacy Center, Center for Marine Conservation, San Francisco Bay Keeper, and a number of other concerned groups to require regulation of vessel ballast water discharges through Clean Water Act permits, arguing that the Coast Guard was the more appropriate regulatory agency.[50] This decision did not alter the development and implementation of the joint EPA-DOD regulations covering discharges from DOD vessels (discussed above) because this DOD effort is specifically authorized in statute. In late December 2003, three Pacific Coast environmental groups filed suit, seeking to force the EPA to regulate ballast water discharges under the Clean Water Act.[51]

On September 18, 2006, the federal district court ruled that EPA's regulation exempting ballast water discharges from the Clean Water Act was contrary to congressional intent and ordered EPA to promulgate new regulations within two years (i.e., September 30, 2008).[52] This ruling, which the government appealed to the U.S. Court of Appeals for the Ninth Circuit,[53] directs EPA to ensure that shipping companies and other vessels comply with the Clean Water Act by restricting the discharge of invasive species in ballast water under that act's permit program.[54] In response, EPA published regulations for vessel general permits (VGPs) on December 29, 2008.[55] Environmental groups expressed concern over the effectiveness of the EPA regulations and filed a court challenge. In a March 2011 settlement, EPA was required to revise its VGP program to require numeric limits on organisms in ballast water discharges as well as additional monitoring and reporting of vessel discharges.[56] Under the terms of this settlement, EPA proposed two new VGP programs on November 30, 2011, with final action on the draft permits anticipated by November 30, 2012.[57]

In light of EPA action, concern arose in the 110th Congress as to what extent legislation proposing national standards for ballast water management to be administered by the Coast Guard might conflict with EPA regulation of ballast water discharges as water pollutants under the Clean Water Act.[58] In response to these concerns, the 110th Congress enacted P.L. 110-299 to provide a two-year moratorium on Clean Water Act VGP permitting for discharges (except for ballast water discharges) from commercial fishing vessels and non-recreational vessels less than 79 feet in length in response to the Northwest Environmental Advocates litigation.[59] During the moratorium, EPA studied the discharges from these vessels and submitted a report to Congress.[60] The 111th Congress, in P.L. 111-215, extended this moratorium through December 18, 2013.

EPA's Science Advisory Board's Ecological Processes and Effects Committee has met several times to provide advice on technologies and systems to minimize the impacts of invasive species in vessel ballast water management, including the issue of whether to support the development of onshore ballast water treatment options. In addition, EPA and Coast Guard have jointly commissioned two studies, including one by the National Research Council, to improve understanding of ballast water discharges.[61]

State Programs

In addition to federal programs and sometimes in response to perceived deficiencies in federal regulation, several states have chosen to regulate aspects of ballast water management, including California,[62] Oregon,[63] Washington,[64] Wisconsin,[65] Minnesota,[66] New York,[67] and Michigan.[68] Particular concerns have arisen over New York State ballast water regulations, called the toughest in the world by some, as to whether the technology exists to meet the standards.[69] In light of state action, several bills in the 110th Congress proposed to preempt state ballast water discharge standards that are more stringent than federal standards, while continuing to allow states to inspect and enforce federal ballast water standards. In the 112th Congress, Section 459 of H.R. 2584, as reported by the House Committee on Appropriations, would prohibit EPA from providing funds to any Great Lakes state that has a more stringent performance or ballast water exchange standard than either a revised Coast Guard standard or the IMO standard.

International Efforts

In July 1991, the International Maritime Organization's (IMO's) Marine Environmental Protection Committee (MEPC) issued voluntary *International Guidelines for Preventing the*

Introduction of Unwanted Aquatic Organisms and Pathogens for Ships'Ballast Water and Sediment Discharges, adopted in Resolution (50) 31 by a diplomatic conference of the IMO in November 1993. IMO members were requested to follow these guidelines, which also called for exchange of ballast water in the open ocean (to reduce transfer of species from port to port). A review conducted by Australia in 1993 revealed that few countries had implemented the guidelines. In 1994, the MEPC established a ballast water working group to draft regulations for the control and management of ships' ballast water. These draft regulations were debated at the November 1998 and June 1999 MEPC meetings. IMO subsequently proposed these management protocols as a formal IMO instrument. This instrument requires all ratifying member nations to follow the regulations, which would include open-ocean exchange.

On February 13, 2004, an *International Convention for the Control and Management of Ships' Ballast Water and Sediments* was adopted at the International Conference on Ballast Water Management for Ships in London, England. This convention will enter into force 12 months after ratification by 30 nations, representing 35% of the world merchant shipping tonnage.[70] The United States was one of the major proponents of this convention, which requires all ships to implement a Ballast Water and Sediments Management Plan. In addition, all ships must carry a Ballast Water Record Book and will be required to conduct BWM procedures in conformity with a specified standard. These standards will be phased in for various vessels, depending upon when they were constructed.[71] Resistance by vessel owners to the new international requirements for ballast water treatment is tempered by their interest in global standards under the IMO convention as opposed to the increasing number of national ballast water programs developed around different approaches for addressing this concern. However, proposed U.S. Coast Guard regulations could mandate ballast water treatment standards as much as 100 times more stringent than those required by the IMO convention, possibly making the IMO convention irrelevant, if the more stringent standards could be met with readily available treatment technology at no substantially greater expense.[72] A Global Industry Alliance—a partnership among IMO, the United Nations Development Program, the Global Environment Facility, and four private shipping corporations—was formed in March 2009 to coordinate international research and development of cost-effective BWM technologies.[73] The United States has neither signed nor has become a contracting party to this IMO convention.

Bilaterally with Canada, the United States is cooperating through the North American Commission for Environmental Cooperation (CEC),[74] the Great Lakes Commission, and the International Joint Commission to better understand, coordinate, and address ballast water management concerns. Improved ballast water management is one component of the CEC's project on "Closing the Pathways of Aquatic Invasive Species across North America."[75] The Great Lakes Commission convened the "Great Lakes Panel on Aquatic Nuisance Species" in late 1991 to coordinate efforts, including ballast water management.[76] In addition, the International Joint Commission (IJC) views invasive species management as an important component of water quality.[77] Furthermore, the Saint Lawrence Seaway Development Corporation jointly administers regulations and seeks to harmonize ballast water regulations for vessels transiting U.S. waters of the Seaway after having operated outside the U.S. EEZ with those required by Canadian authorities for transit in waters under Canadian jurisdiction.[78]

CONGRESSIONAL ACTION

Numerous congressional hearings on ballast water issues and implementation of NISA were held in recent Congresses.[79] Certain testimony at these hearings has been particularly critical of the slow progress in promulgating regulations to implement NISA.

In the 112th Congress, on July 13, 2011, the House Committee on Transportation and Infrastructure, Subcommittees on Coast Guard and Maritime Transportation and on Water Resources and Environment, held a joint hearing on ballast water discharge regulation. Legislation has been introduced in the 112th Congress relating to ballast water:

- Section 5 of S. 1430 would authorize a "green ships" program, with one element focusing on identifying, evaluating, testing, demonstrating, and improving marine technologies for controlling aquatic invasive species; on December 7, 2011, the Senate Committee on Commerce, Science, and Transportation reported (amended) this bill (S.Rept. 112-99).
- H.R. 2840 would amend the Clean Water Act to add a new Section 321 to implement ballast water management and standards related to discharges from commercial vessels; on November 3, 2011, the House Committee on Transportation and Infrastructure reported (amended) this bill (H.Rept. 112-266). On November 4, 2011, the House, by floor amendment, added the language of H.R. 2840 as Title VII of H.R. 2838; the House passed H.R. 2838 (amended) on November 15, 2011.
- Section 459 of H.R. 2584, as reported by the House Committee on Appropriations on July 19, 2011 (H.Rept. 112-151), would have prohibited the Environmental Protection Agency (EPA) from providing funds to any Great Lakes state that has a more stringent performance or ballast water exchange standard than either a revised Coast Guard standard or the International Maritime Organization standard. However, FY2012 appropriations for EPA, included in P.L. 112-74, contain no similar provision.

End Notes

[1] For an in-depth discussion of invasive species generally and their various mechanisms of invasion, see CRS Report RL30123, Invasive Non-Native Species: Background and Issues for Congress, by M. Lynne Corn et al. Zebra mussels and other species arriving in ballast water are specifically discussed in "A Gallery of Harmful Non-Native Plants and Animals" in that report.

[2] Colonies of zebra mussels accumulate and block water-intake pipes and screens of drinking water facilities, industrial facilities, power-generating plants, golf course irrigation pipes, and cooling systems of boat engines.

[3] Information from the Natural Resources Defense Council webpage at http://www. nrdc.org/ greengate/wildlife/ invasivef.asp.

[4] Fred C. Dobbs and Andrew Rogerson, "Ridding Ships' Ballast Water of Microorganisms," Environmental Science & Technology, v. 39, no. 12 (June 15, 2005): 259A-264A.

[5] G. Tracy Mehan, Assistant Administrator for Water, U.S. Environmental Protection Agency, testimony before the Senate Committee on Environment and Public Works, Subcommittee on Fisheries, Wildlife, and Water, June 17, 2003.

[6] 16 U.S.C. §4701(a)(4).

[7] Title I of P.L. 101-646; 16 U.S.C. §§4701-4741.

[8] Massachusetts Institute of Technology Sea Grant, Marine Bioinvasions Fact Sheet: Ballast Water, Cambridge, MA, December 4, 2002; available at http://massbay.mit.edu/ exoticspecies/ballast/fact.html.

[9] National Ballast Information Clearinghouse, Present Ballast Water Management Practices; available at http://invasions.si.edu/nbic/managementpract.html.

[10] A detailed discussion and evaluation of various ballast water treatment options is presented in Chapter 4 ("Shipboard Treatment Options") of Commission on Engineering and Technical Systems, Stemming the Tide: Controlling Introductions of Nonindigenous Species by Ships' Ballast Water, National Academy of Sciences (Washington, DC: 1996), 141 pp.

[11] For example, Australian researchers established a pilot plant to sterilize ballast water using a combination of filtration, ultraviolet light, and sonic disintegration. For more information, see David Salt, "Shipboard Pests get Sterile Treatment," ABC Science Online, September 30, 2003; available at http://www.abc.net.au/science/ news/enviro/ EnviroRepublish_ 956770.htm.

[12] The Coast Guard is studying methods to address NOBOB concerns; for more information, see 70 Fed. Reg. 1448-1449 (January 7, 2005).

[13] David T. Stocks, "Biocides Testing in the Great Lakes," BMT Focus (Summer 2002): 12; available at http://www.bmt.org/brochures/Focus%20Summer%202002.pdf.

[14] State Water Resources Control Board, Evaluation of Ballast Water Treatment Technology for Control of Nonindigenous Aquatic Organisms, California Environmental Protection Agency (December 2002); available at http://www.calepa.ca.gov/Publications/ Reports/ Mandated/2002/BallastWater.pdf.

[15] Frans J. Tjallingii, Market Opportunities for Ballast Water Treatment, Royal Haskoning, International Ballast Technology Investment Fair, Chicago, IL, September 21, 2001.

[16] As of October 2010, U.S. flag vessels comprised 0.81% of the global merchant fleet tonnage, according to statistics from U.S. Maritime Administration, available at http://www. marad.dot.gov/library_ landing_page/data_and_statistics/ Data_and_Statistics.htm.

[17] Title I of P.L. 101-646; 16 U.S.C. §§4701, et seq.

[18] 64 Fed. Reg. 26672-26690 (May 17, 1999). The voluntary guideline interim regulations were effective July 1, 1999. Final regulations were published in 66 Fed. Reg. 58381-58393 (November 21, 2001).

[19] If the voluntary program did not result in sufficient compliance, reporting of BWM practices would become mandatory for nearly all vessels entering U.S. waters (33 CFR 151.2040).

[20] For an analysis of some of the data, see A. Whitman Miller et al., "Geographic Limitations and Regional Differences in Ships' Ballast Water Management to Reduce Marine Invasions in the Contiguous United States," BioScience, v. 61, no. 11 (November 2011): 880-887.

[21] G. M. Ruiz, et al., Status and Trends of Ballast Water Management in the United States: First Biennial Report of the National Ballast Information Clearinghouse (Edgewater, MD: Smithsonian Environmental Research Center, November 16, 2001), p. 4.

[22] U.S. Coast Guard, Report to Congress on the Effectiveness of the Voluntary BWM Program, June 3, 2002.

[23] 68 Fed. Reg. 523-530 (January 6, 2003).

[24] 69 Fed. Reg. 32864-32871 (June 14, 2004).

[25] 69 Fed. Reg. 44952-44961 (July 28, 2004).

[26] As defined for the National Estuary Program in 33 U.S.C. §1330.

[27] Letter of February 11, 1999, to Hon. Carol Browner, Administrator of the U.S. Environmental Protection Agency, from Representatives George Miller, Jim Saxton, and 16 other Members of the U.S. House of Representatives.

[28] Union of Concerned Scientists, The National Invasive Species Act: An Information Update (Cambridge, MA: August 2002); available at http://www.ucsusa.org/assets/ documents /invasive_species/nisa-1.pdf.

[29] An April 29, 2008 lawsuit by the Izaak Walton League and several other environmental groups in a Minnesota federal district court seeks to require the Coast Guard to prevent vessels from taking on and moving ballast water from waters of the Great Lakes to prevent the spread of viral hemorrhagic septicemia to Lake Superior from adjacent lakes. For additional information, see http://www.greatlakeslaw.org/blog/files/ vhs_complaint. pdf.

[30] 58 Fed. Reg. 18330-18335 (April 8, 1993). Regulations were extended to include vessels entering the Hudson River in 59 Fed. Reg. 67632-67634 (December 30, 1994).

[31] Personal communication with Jill R. Teixeira, Lieutenant Junior Grade, U.S. Coast Guard, Office of Congressional & Governmental Affairs, Washington, DC, March 30, 2004.

[32] 68 Fed. Reg. 44691-44696 (July 30, 2003).

[33] 69 Fed. Reg. 44952-44961 (July 28, 2004).

[34] 72 Fed. Reg. 2536-2537 (January 19, 2007).

[35] 67 Fed. Reg. 9632-9638 (March 4, 2002).

[36] 68 Fed. Reg. 55559-55563 (September 26, 2003).

[37] 71 Fed. Reg. 25798-25799 (May 2, 2006).

[38] 74 Fed. Reg. 44632-44672 (August 28, 2009).

[39] 77 Fed. Reg. 17254-17320 (March 23, 2012). Updated Coast Guard information is available at http://www.uscg.mil/ hq/cg5/cg522/cg5224/bwm.asp.

[40] Descriptive material on the Coast Guard proposal is available at http://www.uscg.mil/hq /cg5/cg522/cg5224/docs/ White%20Paper%20-%20Ballast%20Water% 20Discharge% 20Standard%20v3B.pdf.

[41] For additional information, see http://www.uscg.mil/hq/g-m/nvic/04/NVIC_01-04.pdf.

[42] 69 Fed. Reg. 47453-47454 (August 5, 2004). A generic protocol for verification of ballast water treatment technology is published at http://www.epa.gov/nrmrl/pubs/ 600r10146/ 600r10146.pdf.

[43] 74 Fed. Reg. 44632-44672 (August 28, 2009).

[44] Available at http://www.epa.gov/nrmrl/pubs/600r10146/600r10146.pdf.

[45] 70 Fed. Reg. 51831-51836 (August 31, 2005).

[46] The last request for proposals for ballast water treatment technology testing and demonstration projects was published by NOAA at 71 Fed. Reg.33898-33929 (June 12, 2006).

[47] A notice of grant funding available for FY2008 was published at 72 Fed. Reg. 73325-73335 (December 27, 2007).

[48] For more information on the NBIC, see http://invasions.si.edu/NBIC/ballast.html.

[49] 33 U.S.C. §§1251, et seq.

[50] 68 Fed. Reg. 53165-53166 (September 9, 2003). A copy of the EPA decision was available at http://www.epa.gov/ npdes/pubs/ballast_report_petition_response.pdf.

[51] Cathy Zollo, "Environmental Groups Seek EPA Rules for Ballast Water," Naples Daily News, December 24, 2003, available at http://www.naplesnews.com/news/2003/dec/ 24/ndn_ environmental_groups_seek_epa_ rules_for_ballas/.

[52] Northwest Environmental Advocates, et al. v. U.S. Environmental Protection Agency, Case No. C 03-05760 SI (N.D. Cal., decided September 18, 2006).

[53] The appellate court upheld the lower court decision; 9th Circuit Court, on appeal from the U.S. District Court for the District of Northern California, July 23, 2008.

[54] EPA estimated that approximately 61,000 domestically flagged commercial vessels and approximately 8,000 foreign flagged vessels may be affected by this permit.

[55] 73 Fed. Reg. 79473-79481; a fact sheet is available at http://www.epa.gov/npdes/ pubs/vgp_factsheet_ english.pdf.

[56] See http://blogs.law.stanford.edu/newsfeed/files/2011/03/03-08-11-BallastWaterSettlement FINAL.pdf.

[57] For detailed information on the EPA's proposed permits, see CRS Report R42142, EPA's Proposed Vessel General Permits: Background and Issues, by Claudia Copeland.

[58] Detailed information on legislation addressing BWM issues in the 110th Congress can be found in CRS Report RL34640, Regulating Ballast Water Discharges: Legislative Issues in the 110th Congress, by Claudia Copeland.

[59] See CRS Report RS22878, Clean Water Act: 110th Congress Legislation on Discharges from Recreational Boats.

[60] EPA, Report to Congress: Study of Discharges Incidental to Normal Operation of Commercial Fishing Vessels and Other Non-Recreational Vessels Less Than 79 Feet; available at http://www.epa.gov/npdes/pubs/ vesselsreporttocongress_all_final.pdf.

[61] Links to copies of these studies and supporting documentation are available at http://cfpub.epa.gov/npdes/vessels/ programdevelopment.cfm.

[62] See http://www.slc.ca.gov/division_pages/depm/depm_programs_and_reports/ shore_ terminals/text/ 40%20appendix%20e.doc for details on California's program.

[63] See http://www.deq.state.or.us/lq/cu/emergency/ballast.htm for details on Oregon's program.

[64] See http://wdfw.wa.gov/ais/ballast/ for details on Washington's program.

[65] See http://dnr.wi.gov/news/BreakingNews_Lookup.asp?id=1500 for details on Wisconsin's program.

[66] See http://www.pca.state.mn.us/index.php/water/water-permits-and-rules/water-permits-and-forms/vessel-dischargeballast-water-program.html?menuid=&redirect=1 for details on Minnesota's program.

[67] See http://www.dec.ny.gov/permits/72399.html for details on New York's program.

[68] See http://www.michigan.gov/deq/0,1607,7-135-3313_3682_3713-153446—,00.html for details on Michigan's program.

[69] For example, see http://www.northcountrypublicradio.org/news/story/18090/20110727/new-york-s-tough-ballast-water-rules and http://www.jsonline.com/news/wisconsin/ 129777183. html.

[70] As of March 2012, this convention had been ratified by 33 nations, representing 26.46% of the world merchant shipping tonnage. Updated status information can be found at http://www. imo.org/OurWork/Environment/ BallastWaterManagement/Pages/Default.aspx#5.

[71] Details on this new convention were available at http://globallast.imo.org/index.asp? page=mepc.htm&menu= true.

[72] Rajesh Joshi, "New US Ballast Water Plan a Threat to IMO Convention," Lloyd's List, May 1, 2008.

[73] IMO, United Nations Agencies Partner with Shipping Corporations to Minimize Global Environment Threat, IMO/UNDP Joint Press Release, March 2, 2009.

[74] Established under the authority of the North American Free Trade Agreement.

[75] Additional information on this project was available at http://www.cec.org/programs_ projects/conserv_biodiv/ project/index.cfm?projectID=20&varlan=english.

[76] Additional information on this panel was available at http://www.glc.org/ans/.

[77] Additional information on these concerns was available from the IJC's annual report, available at http://www.ijc.org/ php/publications/html/11br/english/report/chapter3/.

[78] 73 Fed. Reg. 9950-9954 (February 25, 2008).

[79] For example, U.S. Congress, Committee on Resources, Subcommittee on Fisheries Conservation, Wildlife, and Oceans and Committee on Science, Subcommittee on Environment, Technology, and Standards, H.R. 5395 and H.R. 5396, Joint legislative hearing, 107th Cong., 2d sess., (Washington, DC: November 14, 2002), 80 pp.; U.S. Congress, Committee on Resources, Subcommittee on Fisheries Conservation, Wildlife, and Oceans, and Subcommittee on National Parks, Recreation, and Public Lands, The Growing Problem of Invasive Species, joint legislative hearing, 108th Cong., 1st sess. (Washington, DC: April 29, 2003), 133 pp.; U.S. Congress, Committee on Environment and Public Works, Subcommittee on Fisheries, Wildlife, and Water, National Aquatic Invasive Species Act of 2003, legislative hearing, 108th Cong, 1st sess. (Washington, DC: June 17, 2003), 240 pp.; an as-yet-unpublished joint hearing of March 25, 2004, by the House Committee on Transportation and Infrastructure, Subcommittees on Coast Guard and Maritime Transportation and on Water Resources and Environment, on Ballast Water Management: New International Standards and National Invasive Species Act Reauthorization; U.S. Congress, Committee on Government Reform, Subcommittee on Regulatory Affairs, Protecting Our Great Lakes: Ballast Water and the Impact of Invasive Species, oversight field hearing, 109th Congress, 1st sess. (Fair Haven, MI: September 9, 2005), 133 pp.; an as-yet-unpublished legislative hearing of June 15, 2005, by the Senate Committee on Commerce, Science, and Transportation, National Ocean Policy Study, on Ballast Water Invasive Species Management and Threats to Coral Reefs; U.S. Congress, Committee on Transportation and Infrastructure, Subcommittee on Water Resources and Environment, The Impact of Aquatic Invasive Species on the Great Lakes, legislative hearing, 110th Congress, 1st sess. (Washington, DC: March 7, 2007), 256 pp.

In: Ballast Water and Invasive Species
Editor: Gilles Wuyts

ISBN: 978-1-62618-754-2

Chapter 2

EFFICACY OF BALLAST WATER TREATMENT SYSTEMS: A REPORT BY THE EPA SCIENCE ADVISORY BOARD*

EPA Science Advisory Board

1. EXECUTIVE SUMMARY

Vessel ballast water discharges are a primary source of introductions of nonindigenous species and potentially harmful pathogens to marine, estuarine, and freshwater ecosystems of the United States. At present, federal requirements for managing ballast water discharges rely primarily on ballast water exchange. However, the U.S. Coast Guard (USCG) has proposed numeric concentration-based limits for live organisms in ballast water that would initially require compliance with a "Phase 1 standard" that has the same concentration limits as the International Maritime Organization (IMO) D-2 standard and subsequently require compliance with a more stringent "Phase 2 standard." In addition, the U.S. Environmental Protection Agency is considering numeric standards that limit the number of live organisms in discharged ballast water in its revision of the Vessel General Permit.

This Advisory report responds to a request from the EPA Office of Water (OW) for EPA's Science Advisory Board (SAB) to provide advice on technologies and systems to minimize the impacts of invasive species in vessel ballast water discharge. More specifically, the SAB was requested to provide review and advice regarding whether existing shipboard treatment technologies can reach specified concentrations of organisms in vessel ballast water, how these technologies might be improved in the future, and how to overcome limitations in existing data. To conduct the assessment, the SAB's Ecological Processes and Effects Committee (EPEC) was augmented with additional Panel members (Ballast Water Advisory Panel, or "Panel") having expertise in ballast water issues, marine engineering, and engineering treatment technologies.

* This is an edited, reformatted and augmented version of the U.S. Environmental Protection Agency, EPA-SAB-11-009, dated July 12, 2011.

In addition to the SAB assessment of ballast water management systems (BWMS), the EPA and the USCG commissioned the National Academy of Sciences (NAS) to conduct a complementary study[1] that assesses the risk of successful invasions as a function of different concentrations of organisms in ballast water discharges. Therefore, this SAB Advisory does not address the relationship between the various concentrations of organisms as described in proposed standards and the likelihood of invasions or the potential effects of pathogens. Rather, in this report, “effectiveness” refers to how well a technology may meet a given numeric concentration limit for organisms, not to how well the technology may protect the environment. Similarly, phrases such as “meets more stringent” or “meets less stringent” proposed standards are used as descriptive conclusions and do not represent performance recommendations.

To prepare this Advisory report, the Panel reviewed a “Background and Issue Paper” written by EPA’s OW and USCG (Albert et al. 2010). This paper provided an overview of information about major categories of shipboard ballast water treatment technologies and presented proposed ballast water discharge standards drawn from international sources, the USCG, and nine states. In addition, EPA’s OW and the public identified information on 51 existing or developmental ballast water treatment technologies, although detailed data were available for only 15 specific BWMS. The Panel used this information as the source material for its assessment of ballast water treatment performance and, as requested by the EPA, used proposed ballast water discharge standards as the performance benchmarks.

Regulatory Context

Ballast water discharges are regulated by EPA under authority of the Clean Water Act (CWA) and by the USCG under authority of the National Invasive Species Act (NISA). In December 2008, EPA issued a Vessel General Permit (VGP) for discharges incidental to the normal operation of commercial vessels, including ballast water discharges. The VGP sets effluent limits for ballast water that rely on “best management practices” (primarily use of ballast water exchange, or BWE) and do not include a numeric discharge limit. The VGP will expire on Dec. 19, 2013. For subsequent iterations of the VGP, the EPA has stated its intention to establish best available technology standards for the treatment of ballast water, once such technologies are shown to be commercially available and economically achievable.

Existing USCG rules governing ballast water also primarily rely on BWE. In August 2009, the USCG proposed revisions to their existing rules to establish numeric concentration-based limits for viable organisms in ballast water. The proposed USCG rule would initially require compliance with a “Phase 1” standard, and, if a practicability review shows it is feasible, it would be followed by a “Phase 2” standard that sets concentration limits at 1000 times more stringent than Phase 1 standards for viable organisms >10 μm in minimum dimension. Phase 2 standards also set limits on the discharge concentration for bacteria and viruses. Neither Phase 1 nor Phase 2 standards have been finalized. The USCG Phase 1 standards have essentially the same concentration limits as those adopted in 2004 by the International Maritime Organization’s (IMO) International Convention for the Control and Management of Ships’ Ballast Water and Sediments; thus both standards are often referred to as the “D-2/ Phase 1 standards.” The U.S. is not a Party to the Convention, nor has the

Convention yet entered into force. However, manufacturers of BWMS have generally designed their equipment to meet these IMO D-2 standards.

Ballast Water Management Should be Implemented Using a Risk-Based Systems Approach

The Panel recommends that any ballast water management strategy to decrease the rate of successful invasions by nonindigenous species or introduction of pathogens be part of an overall risk-based management plan that includes methods to reduce invasion events, process and environmental monitoring, containment, and eradication. Emphasis only on one aspect, the initial introduction of organisms, is not likely to reduce the risk of invasions as effectively or as cost-efficiently as a risk assessment approach that considers all the stages of the invasion process including survival after introduction. Decisions on approaches to ballast water management should be viewed in the context of risk management and should: (1) recognize the stochastic and non-linear nature of the invasion process, (2) clearly define the management goals, and (3) evaluate the effectiveness of BWMS within the context of other sources of nonindigenous species and other organisms found on the vessel and in the treatment system, and with respect to specific receiving habitats.

Rigorous Sampling and Statistical Verification of Performance is Essential

The Panel was asked to respond to charge questions that focused primarily on whether test data demonstrated that BWMS met or "closely approached" proposed standards for discharge and whether they did so "credibly" and "reliably." As benchmarks for performance, the Panel was asked to consider proposed numerical standards as well as narrative descriptions such as "no living organisms," "sterilization," and "zero or near zero" discharge. In order to place its assessments of treatment performance in appropriate scientific context, the Panel first had to consider statistical and sampling issues. While "zero detectable discharge" might initially seem a desirable standard, it is not statistically verifiable. Further, verification of standards that set very low organism concentrations may require water samples that are too large to be logistically feasible. However, when small sample volumes are used, the probability of detecting an organism is low even when the actual organism concentration is relatively high. These errors depend on the sample volume collected, and the relative errors are much larger for small sample volumes.

A well-defined, rigorous sampling protocol is necessary to assess the effectiveness of BWMS at meeting different standards. These sampling protocols should include consideration of the spatial distribution of plankton in ballast water. The Poisson distribution is recommended as the model for statistical analysis of treated water samples.

The Panel also concludes that the D-2/Phase 1 performance standards for discharge quality are currently measurable, based on data from land-based and shipboard testing. However, current methods (and associated detection limits) prevent testing of BWMS to any standard more stringent than D-2/Phase 1 and make it impracticable for verifying a standard 100 or 1000 times more stringent. New or improved methods will be required to increase detection limits sufficiently to statistically evaluate a standard 10x more stringent than IMO

D-2/Phase 1; such methods may be available in the near future. These conclusions pertain to evaluating data from land-based and shipboard testing, although the same statistical theory and practice applies to compliance testing by port state control officers.

Charge Question 1: Performance of Shipboard Systems with Available Effluent Testing Data

a. For the shipboard systems with available test data, which have been evaluated with sufficient rigor to permit a credible assessment of performance capabilities in terms of effluent concentrations achieved (living organisms/unit of ballast water discharged or other metric)?

Evaluations of technologies are necessarily based on performance information for a given point in time and the development and manufacture of ballast water treatment systems is a dynamic industry. For this assessment, the Panel reviewed information provided by EPA's Office of Water and the public. This information included peer-reviewed articles and publications; information provided directly from individual manufacturers of BWMS (some included data reports, others provided only Type Approval certificates); and public dossiers submitted to the IMO Group of Experts on the Scientific Aspects of Marine Environmental Protection (GESAMP). This information was prepared or published prior to May, 2010. However, the majority of the documents were from 2008 to 2010, reflecting growth in the BWMS industry. While other BWMS may exist, the Panel considered only those for which information was provided.

From this information, the Panel identified 51 individual BWMS, which can be grouped into 34 categories of treatment technologies. Of the 51 BWMS identified, the Panel concluded that test data and other information for 15 individual BWMS were credible and sufficient to permit an assessment of performance capabilities. Of these 15 BWMS, nine systems (representing individual configurations of five different categories of BWMS) achieved significant reductions in organism concentrations, and were able to comply with the Phase 1 standard. These five categories of BWMS technologies are: (1) Deoxygenation + cavitation; (2) Filtration + chlorine dioxide; (3) Filtration + UV; (4) Filtration + UV + TiO_2; and (5) Filtration + electro-chlorination.

b. For those systems identified in (1a), what are the discharge standards that the available data credibly demonstrate can be reliably achieved? Furthermore, do data indicate that certain systems (as tested) will not be able to reliably reach any or all of the discharge standards shown in that table?

The Panel concluded that the same five BWMS categories (listed above) have been demonstrated to meet the IMO D-2 discharge standard, when tested under the IMO certification guidelines, and will likely meet USCG Phase 1 standards, if tested under EPA's more detailed Environmental Technology Verification (ETV) Protocol (U.S. EPA 2010). The Panel acknowledges the significant achievement of several existing BWMS to effectively and reliably remove living organisms from ballast water under the challenging conditions found on active vessels.

The detection limits for currently available test methods preclude a complete statistical assessment of whether BWMS can meet standards more stringent than IMO D-2/Phase 1. However, based on the available testing data, it is clear that while five types of BWMS are able to reach IMO D-2/Phase 1, none of the systems evaluated by the Panel performed at 100 times or 1000 times the Phase 1 standard.

c. For those systems identified in (1a), if any of the system tests detected "no living organisms" in any or all of their replicates, is it reasonable to assume the systems are able to reliably meet or closely approach a "no living organism" standard based on their engineering design and treatment processes?

The Panel concluded that it is not reasonable to assume that BWMS are able to reliably meet or closely approach a "no living organism" standard. Available data demonstrate that current BWMS do not achieve sterilization or the complete removal of all living organisms.

Charge Question 2: Potential Performance of Shipboard Systems without Reliable Testing Data

Based on engineering design and treatment processes used, and shipboard conditions/constraints, what types of ballast water treatment systems can reasonably be expected to reliably achieve any of the proposed standards, and if so, by what dates? Based on engineering design and treatment processes used, are there systems which conceptually would have difficulty meeting any or all of the proposed discharge standards?

The Panel found that nearly all of the 51 BWMS evaluated are based on reasonable engineering designs and treatment processes, and most are adapted from long-standing water treatment approaches. However, the lack of detailed information on the great majority of BWMS precludes an assessment of limitations in meeting any or all discharge standards. In particular, the Panel determined that the following data are essential to future assessments: documentation that test protocols were followed; full reporting of all test results; and documentation that rigorous QA/QC methods were followed.

Although several BWMS appear to safely and effectively meet IMO D-2/Phase 1 discharge standards, the Panel notes that factors beyond mechanical and biological efficacy need to be considered as BWMS technology matures. Several parameters will affect the performance or applicability of individual BWMS to the wide variety of vessel types that carry ballast water. These include environmental parameters (e.g., temperature and salinity), operational parameters (e.g., ballast volumes and holding times), and vessel design characteristics (e.g., ballast volume and unmanned barges).

Charge Question 3: System Development

a. For those systems identified in questions 1 a. and 2, are there reasonable changes or additions to their treatment processes which can be made to the systems to improve performance?

The Panel defined "reasonable changes" as moderate adjustments that do not fundamentally alter the treatment process. Based on information from available test results, such moderate adjustment could be made to treatment processes, although it may add costs and engineering complexity. Examples of moderate adjustments are:

- Deoxygenation + cavitation. It may be possible to reduce the time needed to reach severe hypoxia, to increase holding time under severe hypoxia, and to increase the degree of cavitation and physical/mechanical disruption of organisms.
- Mechanical separation + oxidizing agent. These systems could be optimized by improving mechanical separation, increasing concentration and contact time for oxidizing agents, and adjusting other water chemistry parameters (e.g., pH) to increase oxidizing agent efficacy.
- Mechanical separation + UV. These systems could be optimized by improved mechanical separation and by increasing UV contact time and dosage.

The Panel concludes that moderate adjustments or changes to existing combination technologies are expected to result in only incremental improvements. Reaching the Phase 2 standard, or even 100x IMO D-2/ Phase1, would require wholly new treatment systems. Such new systems likely would use new technological devices, including those drawn from the water treatment industry; employ multistage treatment processes; emphasize technological process controls and multiple monitoring points; include physical barriers to minimize the potential for cross-contamination of the system; and become part of an integrated ballast water management effort. These new approaches likely will achieve higher performance, but will require time to develop and test in order to determine their practicality and cost.

b. What are the principal technological constraints or other impediments to the development of ballast water treatment technologies for use onboard vessels to reliably meet any or all of the discharge standards?

Existing BWMS have been developed within the context of typical marine vessel constraints, including restrictions on size, weight, and energy demands. The primary impediments to the ability of shipboard systems to meet stringent discharge standards are that treatment processing plants likely will be large, heavy, and energy intensive—many existing vessels may be unable to overcome these barriers through retrofitting BWMS. Meeting more stringent performance standards may require a fundamental shift in how ballast water is managed. Existing and potential BWMS share several common impediments to development: (1) The focus has been on engineering the technology with less attention to equally important issues such as training, operation, maintenance, repair, and monitoring. (2) Without an established compliance monitoring and enforcement regime to guide design requirements for technologies, incentives for further innovations are dampened. (3) Facilities properly equipped to test BWMS technologies are few, so increased sharing of data and testing protocols among such facilities is essential. (4) Discharge standards differ domestically and internationally, giving manufacturers multiple targets. (5) Meeting more stringent standards requires that BWMS consistently perform nearly perfectly; a fundamental shift in system design and operational practices would be needed to achieve this level of performance. (6)

Once performance tests indicate that a given BWMS meets IMO D-2/ Phase 1 standards, further efforts by manufacturers to improve design and efficacy appear to decline.

c. What recommendations does the SAB have for addressing these impediments and constraints?

Clearly defined and transparent programs for compliance monitoring and enforcement are needed to promote consistent, reliable operation of BWMS; such programs do not yet exist. Ideally, vessel crew would have the technological capability to self-monitor BWMS efficacy and make real-time corrections to maintain compliance. BWMS manufacturers should document performance metrics beyond discharge treatment efficacy, such as energy consumption and reliability. This would enable vessel operators to select systems that best integrate with their operations. Although meeting significantly higher standards will likely require completely new treatment approaches, the Panel can neither predict which combination of treatment processes will achieve the highest efficacy nor their ultimate performance. The Panel recommends that one or more pilot projects be commissioned to explore new approaches to ballast water treatment, including tests of ballast water transfer and treatment at an onshore reception facility.

d. Are these impediments more significant for certain size classes or types of organisms (e.g., zooplankton versus viruses)? Can currently available treatment processes reliably achieve sterilization (no living organisms or viable viruses) of ballast water onboard vessels or, at a minimum, achieve zero or near zero discharge for certain organism size classes or types?

Shipboard impediments apply to all size classes of organisms and specified microbes. Some treatment systems or combinations are more effective for treating larger organisms and others for treating unicellular organisms. The technology exists to remove or kill the great majority and in some cases, to remove nearly all organisms ≥ 50 μm from discharged water. Given the volumes of water involved, onboard sterilization of ballast water is not possible using current technologies. It is not possible to verify zero (sterilization) or near-zero discharge. Such values cannot be measured in a scientifically defensible way.

Charge Question 4: Development of Reliable Information

What are the principal limitations of the available studies and reports on the status of ballast water treatment technologies and system performance and how can these limitations be overcome or corrected in future assessments of the availability of technology for treating ballast water onboard vessels?

Existing information about ballast water treatment is limited in many respects, including significant limitations in data quality, shortcomings in current methods for testing BWMS and reporting results, issues related to setting standards and for compliance monitoring, and issues related to test protocols, including the use of surrogate indicators. More broadly, however, the Panel found that because of the lack of an overall risk management systems approach, insufficient attention has been given to integrated sets of practices and technologies to reduce invasion or pathogen risk by (1) managing ballast, (2) adjustments in operation and ship design to reduce or eliminate the need for ballast water, (3) development of voyage-based risk assessments and application of Hazard Analysis and Critical Control Points (HACCP) principles, and (4) options for reception facilities for the treatment of ballast water. The Panel concludes that combinations of practices and technologies are potentially more effective and cost-efficient than sole reliance on shipboard BWMS.

Principal Limitations of Available Data and Protocols

Data are not sufficiently compatible to compare rigorously across BWMS because standard test protocols have been lacking. The most recent EPA ETV Protocol, published in 2010, will improve this. Reporting of test failures during type approval testing is not required, although some independent test facilities do report failures. This should be uniform across research and other test facilities so that it is possible to draw conclusions about the consistency or reliability of BWMS. Clear definitions and direct methods to enumerate viable organisms are missing for some organisms and are logistically problematic for all size classes, especially nonculturable bacteria, viruses, and resting stages of many other taxa. Methods to enumerate viruses are not included in the proposed USCG Phase 2 standard. The important size class of protists[2] < 10 μm have not been considered adequately in developing guidelines and standards, although some Panel members felt that other measurements may indicate activity in that size class.

Alternatives to Shipboard Treatment of Ballast Water

Data on the effectiveness of practices and technologies other than shipboard BWMS are few. Use of reception facilities for the treatment of ballast water appears to be technically feasible (given generations of successful water treatment and sewage treatment technologies), and is likely to be more reliable and more readily adaptable than shipboard treatment. Existing regional economic studies suggest that treating ballast water in reception facilities would be at least as economically feasible as shipboard treatment. However, these studies consider that vessels only call at those regional facilities; if vessels also call at ports outside the region without reception facilities, they would need a shipboard BWMS[3]. The effort and cost of monitoring and enforcement needed to achieve a given level of compliance is likely to be less for a smaller number of reception facilities compared to a larger number of BWMS. However, the Panel did not reach agreement on several issues related to treatment of ballast water in onshore reception facilities. For further details on the Panel's views, see Section 6.4; for a review of the literature review on treatment of ballast water in onshore reception facilities, see Appendix B.

Recommendations to Overcome Present Limitations

As illustrated in the ETV protocol (U.S. EPA 2010), testing of BWMS in a research and development mode should be distinct from testing for type approval certification and for verification. Certification testing should be conducted by a party independent from the manufacturer with appropriate, established credentials, approved by EPA/USCG. Test failures and successes during type approval testing should be reported and considered in certification decisions. A transparent international standard format for reporting, including specification of quality assurance and quality control (QA/QC) protocols and a means to indicate QA/QC procedures were followed during testing, are needed. In addition, the EPA should develop metrics and methods appropriate for compliance monitoring and enforcement as soon as possible.

Limits for selected protists < 10 µm in minimum dimension should be included in ballast water discharge standards and in BWMS test protocols. Suitable standard test organisms should be identified for bench-scale testing, and surrogate parameters should be investigated to complement or replace metrics that are logistically difficult or infeasible for estimating directly the concentration of living organisms. Representative "indicator" taxa (toxic strains of *Vibrio cholerae*; *Escherichia coli*; intestinal Enterococci) should continue to be used to assess BWMS. Estimates of the removal of harmful bacteria will be improved when reliable techniques become available to account for active nonculturable cells as well as culturable cells.

EPA should conduct a comprehensive analysis comparing biological effectiveness, cost, logistics, operations, and safety associated with both shipboard BWMS and reception facilities. If the analysis indicates that treatment at reception facilities is both economically and logistically feasible and is more effective than shipboard treatment systems, it should be used as the basis for assessing the ability of available technologies to remove, kill, or inactivate living organisms to meet a given discharge standard. In other words, use of reception facilities may enable ballast water discharges to meet a stricter standard.

A risk management systems approach should be adopted, in which combinations of practices and technologies should be considered as potentially more effective and potentially more cost-efficient approaches than reliance on one ballast water treatment technology. Hazard Analysis and Critical Control Points (HACCP) has been demonstrated to be an effective risk management tool in a variety of situations and could be applied to ballast water management.

2. INTRODUCTION

2.1. EPA's Charge to the SAB

EPA's Office of Water provided the following background and charge questions to the SAB:

Ballast water is typically drawn in from surrounding ambient water and used to assist with vessel draft, buoyancy, and stability. Almost all large vessels have ballast tanks dedicated to this purpose; some vessels may also ballast empty cargo holds. The ballast water discharge rate and constituent concentrations of ballast water from vessels will vary by vessel type, ballast tank capacity, and type of deballasting equipment.

Under current U.S. regulation and permitting requirements (discussed in greater detail in the White Paper), there are existing best management practices to reduce the potential impacts of ballast water discharges. These include ballast water exchange and salt water flushing (collectively referred to as BWE).

While useful in reducing the presence of potentially invasive organisms in ballast water, BWE can have variable effectiveness and may not always be feasible due to vessel safety concerns. In order to make progress beyond use of BWE, establishing a standard for the concentration of living organisms in ballast water that can be discharged is necessary. The United States Environmental Protection Agency (EPA) and the United States Coast Guard (USCG) both desire a stronger federal ballast water management program.

To help develop the next Clean Water Act Vessel General Permit (VGP), EPA needs an objective evaluation of the status and efficacy of ballast water treatment technologies and systems that are in existence or in the development process. A second major scientific question for regulatory agencies is to better understand and relate the concentration of living organisms in ballast water discharges to the probability of introduced organisms successfully establishing populations in U.S. waters.

Given the complexity of the issue, EPA's Office of Water is seeking advice from the Science Advisory Board (SAB) on the first issue and the National Academy of Sciences' National Research Council (NRC) on the second issue. In particular, EPA is seeking advice from the SAB regarding the availability and efficacy of ballast water treatment systems in neutralizing (killing) living organisms that might be discharged from ballast tanks.

For the other NRC study, EPA has requested that the NRC broadly assess and make recommendations about various approaches for assessing the risk of establishment of new aquatic non- indigenous species from ballast water discharges (see attachment 2 of the White Paper for the NRC charge).

Charge Question 1: *Performance of shipboard systems with available effluent testing data*[4]

a. *For the shipboard systems with available test data, which have been evaluated with sufficient rigor to permit a credible assessment of performance capabilities in terms of effluent concentrations achieved (living organisms/unit of ballast water discharged or other metric)?*
b. *For those systems identified in (1a), what are the discharge standards that the available data credibly demonstrate can be reliably achieved (e.g., any or all of the standards shown in Table 1 of the White Paper? Furthermore, do data indicate that certain systems (as tested) will not be able to reliably reach any or all of the discharge standards shown in that table?*
c. *For those systems identified in (1a), if any of the system tests detected "no living organisms" in any or all of their replicates, is it reasonable to assume the systems are able to reliably meet or closely approach a "no living organism" standard or other standards identified in Table 1 of the White Paper, based on their engineering design and treatment processes?*

Charge question 2: Potential performance of shipboard systems without reliable testing data

Based on engineering design and treatment processes used, and shipboard conditions/constraints, what types of ballast water treatment systems (which may include any or all the systems listed in Table 4 of the White Paper) can reasonably be expected to reliably achieve any of the standards shown in Table 1 of the White Paper, and if so, by what dates? Based on engineering design and treatment processes used, are there systems which conceptually would have difficulty meeting any or all of the discharge standards in Table 1 of the White Paper?

Charge question 3: System development

- a. *For those systems identified in questions 1 a and 2, are there reasonable changes or additions to their treatment processes which can be made to the systems to improve performance?*
- b. *What are the principal technological constraints or other impediments to the development of ballast water treatment technologies for use onboard vessels to reliably meet any or all of the discharge standards presented in Table 1 of the White Paper?*
- c. *What recommendations does the SAB have for addressing these impediments /constraints?*
- d. *Are these impediments more significant for certain size classes or types of organisms (e.g., zooplankton versus viruses)?*
- e. *Can currently available treatment processes reliably achieve sterilization (no living organisms or viable viruses) of ballast water onboard vessels or, at a minimum, achieve zero or near zero discharge for certain organism size classes or types?*

Charge question 4: Development of reliable information

What are the principal limitations of the available studies and reports on the status of ballast water treatment technologies and system performance and how can these limitations be overcome or corrected in future assessments of the availability of technology for treating ballast water onboard vessels?

2.2. SAB's Review Process

In response to the EPA request for an SAB assessment of shipboard ballast water management systems (BWMS), the SAB's Ecological Processes and Effects Committee was augmented with additional experts in ballast water issues, marine engineering, and engineering treatment technologies (the Panel). The Panel met on July 29 – 30, 2010 to receive briefings from EPA's OW, to hear public comments, and to begin discussing the charge questions. As requested, the SAB based its assessment and advice on information provided by OW, including the background document, *Availability and Efficacy of Ballast Water Treatment Technology: Background and Issue Paper* (June 2010), and a compilation of information and data on BWMS (described in Appendix A and Section 4 of this report).

Teleconferences were held on October 26 and November 4, 2010, to discuss preliminary texts prepared by individual subgroups of the Panel and to hear public comments. The full Panel met again on January 25-26, 2011, to discuss the compiled draft report and to discuss preliminary conclusions and recommendations. Additional teleconferences were held on March 15 and March 17, 2011, to discuss final revisions to the draft report. The Panel considered public comments provided throughout the advisory process. The Panel differed in their views regarding several issues related to onshore treatment of ballast water in reception facilities. The potential use of reception facilities is discussed in Section 6.4, and a review of the literature of onshore treatment in reception facilities is provided in Appendix B. One panel member did not concur with the final draft report prepared for quality review by the chartered SAB. Public comments were received and considered throughout the advisory process. The chartered SAB conducted a quality review on June 16, 2011, and approved the report with clarifying edits.

2.3. Regulatory Frameworks for Ballast Water Management

2.3.1. U.S. Federal Rules

In December 2008, the EPA issued a Vessel General Permit (VGP) for discharges incidental to the normal operation of commercial vessels, including ballast water, as authorized under the Clean Water Act (CWA). The VGP set technology-based effluent limits for ballast water that rely on "best management practices" and do not include a numeric discharge limit. The required VGP practices include flushing and exchange of ballast water by vessels in Pacific near-shore voyages and saltwater flushing of ballast water tanks that are empty or contain only un-pumpable residual ballast water in addition to mid-ocean exchange. The VGP expires on Dec. 19, 2013.

Existing U.S. Coast Guard (USCG) rules governing ballast water, as authorized under the National Invasive Species Act (NISA), also rely primarily on ballast water exchange. Though the BWE provisions are not identical, the general principle of BWE as used by the EPA and USCG is very similar. NISA generally requires vessels equipped with ballast water tanks and bound for ports or places in the U.S. after operating beyond the U.S. Exclusive Economic Zone either to conduct a mid-ocean ballast water exchange (BWE), retain their ballast water onboard, or use an alternative environmentally sound ballast water management method approved by the USCG. In August 2009, the USCG proposed revising their existing rules to establish numeric concentration-based limits for organisms in ballast water. The proposed rule initially would require compliance with a "Phase 1 standard" that has the same concentration limits as the International Maritime Organization (IMO) D-2 standard (see below) and subsequently would require compliance with a "Phase 2 standard" that is 1000 times (1000x) more stringent for organisms of more than 10 μm in minimum dimension. The Phase 2 standard also contains concentration limits for bacteria and viruses. As of July 2011, the USCG has not finalized this rule, and in the meantime continues to require use of BWE.

In recent years, Congress has considered but not enacted legislation that would directly set concentration-based ballast water discharge standards, including standards that would be 100 times (100x) more stringent for organisms than the USCG proposed Phase 1 standard.

2.3.2. Other Regulatory Frameworks

U.S. States

Under the CWA, U.S. states have the authority to impose their own ballast water discharge standards through the CWA section 401 certification process that applies to federally issued National Pollution Discharge Elimination System (NPDES) permits such as the VGP. A number of states have exercised that authority by setting numeric limits for ballast water discharges into their waters, and these numeric limits are included as a condition in the VGP. In addition, several states (e.g., California and some Great Lakes states) have enacted their own independent state laws to establish ballast water treatment standards. Thus, in practice, EPA's VGP standards establish the minimum standard for ballast water discharges, but states retain and have exercised their authority to set standards that are more stringent.

International Standards / Treaties

The International Convention for the Control and Management of Ships' Ballast Water and Sediments (IMO Ballast Water Management Convention), adopted by the International Maritime Organization (IMO) in February 2004, contains concentration-based limits on organisms in ballast water set out in its Regulation D-2. The treaty will not come into force unless it is ratified by additional countries, and implementation would then require the enactment and enforcement of appropriate laws or regulations by the countries that are party to the treaty. However, equipment manufacturers are currently designing and testing equipment to meet the D-2 standards. These and the other main concentration-based limits for organisms in ballast water that have been proposed or adopted are shown in Table 2-1.

2.3.3. Glossary of Terms Used

To clarify the terms used in this report, the Panel provides the following definitions.

- **BWMS** refers to ballast water management systems as developed by vendors for installation on vessels for treatment of ballast water prior to discharge. In this report, "systems" refers specifically to commercial treatment units, not to a systems approach to ballast water management.

Table 2-1. The range of concentration-based limits proposed or adopted for organisms in ballast water

A. Concentration limits for four organism classes				
	Organisms ≥50 µm in minimum dimension	**Organisms ≥10-<50 µm in minimum dimension**	**Bacteria**	**Viruses**
	per m^3	**per ml**	**per ml**	**per ml**
IMO D-2	10	10	no limit	no limit
USCG Phase 1	10	10	no limit	no limit
US Negotiating Position	0.01	0.01	no limit	no limit
USCG Phase 2	0.01	0.01	10	100
California Interim	no detectable[a]	0.01	10	100
California Final	zero detectable[b]	zero detectable	zero detectable	zero detectable

Table 2-1. (Continued)

B. Public health protective concentration limits			
	Toxicogenic *Vibrio cholerae*	*Escherichia coli*	**Intestinal enterococci**
	per ml	**per ml**	**per ml**
IMO D-2	.01	2.5	1
USCG Phase 1	.01	2.5	1
US Negotiating Position	.01	1.26	0.33
USCG Phase 2	.01	1.26	0.33
California Interim	.01	1.26	0.33
California Final	no detectable	no detectable	no detectable

"US Negotiating Position" is what the U.S. argued for in the negotiations on discharge standards for the IMO Ballast Water Management Convention. "California Interim" and "California Final" standards refer to the limits enacted by the state of California in 2006.

[a] For California's interim standard for organisms ≥ 50 μm, the "no detectable" standard is not associated with a volumetric requirement, i.e., the standard is not "no detectable living organisms" per cubic meter.

[b] California's final standard is set as "zero detectable living organisms for all size classes." This final standard also does not have a volume or organism concentration associated with it.

- **Challenge conditions** refer to the challenge water (influent) conditions as specified by the IMO and the ETV Protocol. The IMO's G8 guidelines for challenge conditions are specified in paragraph 2.2.2.5 (for shipboard testing) and paragraphs 2.3.3 and 2.3.17 – 2.3.22 (for land- based testing) (at http://www.regulations.gov/search/Regs/home.html#documentDetail?R=09000064807e8904). The EPA's Environmental Technology Verification (ETV) draft Generic Protocol for Verification of Ballast Water Treatment Technologies, protocol for challenge conditions are specified in § 5.2. (at http://standards.nsf.org/apps/group_public/download.php/7597/Draft%20ETV%20Ballast%20Water%20Prot-v4%202.pdf).
- **ETV Protocol** refers to the U.S. EPA *Generic Protocol for the Verification of Ballast Water Treatment Technology,* Version 5.1. 2010 (EPA/600/R-10/146), U.S. EPA Environmental Technology Verification Program, Washington, DC.
- **G9 approval, both "Basic Approval" and "Final Approval":** Under Regulation D-3(2) of the IMO Ballast Water Management Convention, ballast water treatment systems that make use of "active substances" (biocides or other potentially harmful substances) are subject to approval by the IMO's Marine Environment Protection Committee (MEPC) with respect to active substance- related health, environmental, and safety issues. This review and approval is conducted under the "G9 Guidelines" developed by MEPC, available at http://www.regulations.gov/search/Regs/home.html# document Detail?R=09000064807e890e.

  Basic Approval" requires laboratory or bench-scale testing, while "Final Approval" requires testing an actual piece of equipment. The Group of Experts on the Scientific Aspects of Marine Pollution (GESAMP), an advisory body established by the United Nations in 1969, conducts the technical reviews and makes approval or denial recommendations to MEPC. MEPC then makes the G9 approval decisions.
- **IMO** refers to the International Maritime Organization, a subsidiary body of the UN whose principal responsibility is to develop and maintain the international regulatory

framework for shipping with respect to safety, environmental concerns, legal matters, technical co-operation, and maritime security. It accomplishes this through treaties negotiated under its auspices, including the February 2004 International Convention for the Control and Management of Ships' Ballast Water and Sediments. The Marine Environment Protection Committee (MEPC) is the principal IMO committee with responsibility for environmental issues associated with shipping. For more information: http://www.imo.org/home.asp

- **IMO D-2** refers to the ballast water discharge standards (expressed as concentrations of organisms per unit of volume) that are contained in Regulation D-2 of the IMO Ballast Water Management Convention. The U.S. has not ratified the treaty, and additional countries must ratify it before it enters into force. Nonetheless, the D-2 standards have had an important influence on the design of shipboard BWMS.
- **IMO D-2/Phase 1** are ballast water discharge standards that are sometimes used in combination because their specifications are very similar. However, to be explicit, **IMO D-2** is defined as shown above.
- **USCG Phase 1** (or **P-1**) refers to ballast water discharge standards contained in the U.S. Coast Guard's August 28, 2009, notice of proposed rulemaking. Because this is a proposed rulemaking that has not yet been finalized, these Phase 1 standards are not currently (as of July 1, 2011) legally binding. For more information, refer to 74 *Federal Register* 44632. The table below contains the text of the standards as stated in IMO D-2 and in the proposed USCG Phase 1, arrayed so as to enable their direct comparison (blanks in table are not omissions, but rather are arranged to highlight comparison of the texts).
- **10x D-2, 100x D-2, 1000x D-2** refer to concentration limits that are 10 times, 100 times, or 1000 times smaller (i.e., more stringent) than the concentration limits specified in IMO D-2, for one or both of the organism size classes in IMO D-2 (i.e., organisms with minimum dimension ≥ 50 μm; or ≥ 10 μm and < 50 μm). The 10x, 100x, and 1000x notations do not apply to the D-2 indicator microorganisms. **100x D-2** has been discussed in other fora such as past Congressional bills and State requirements. **1000x D-2** has been discussed in other fora such as the potential Phase II standards in the USCG August 2009 proposed rule or as described in state requirements
- **Type approval** refers to the process under which a type of equipment is tested and certified by a Flag state or its authorized representative (such as a Class society) as meeting an applicable standard specified in treaty, law or regulation. This testing is conducted on a sample piece of equipment which in all material respects is identical to the follow-on production units. For the IMO Ballast Water Management Convention, type approval testing (sometimes called "efficacy testing") is conducted under the G8 Guidelines described in Regulation D-3(1) of the Convention. The guidelines require both land-based and shipboard testing to verify the equipment's ability to meet the IMO D-2 standards. In the U.S., a generally similar type approval procedure was proposed as part of the USCG's August 28, 2009, notice of proposed rulemaking.
- **Verification Organization (VO)** refers to the party responsible for overseeing the test facility's test Total Quality Assurance Plan (TQAP) development, overseeing

testing activities, and overseeing the development and approval of the Verification Report and Verification Statement for the ballast water treatment system. Within the ETV Program, verification organizations are the managers and operators of the various technology centers under cooperative agreements with the EPA (U.S. EPA 2010).

Table 2-2. Comparing IMO D-2 with USCG Proposed Phase I Standard

IMO Regulation D-2 Standard	USCG Proposed Phase 1 Standards
Discharge less than 10 viable organisms per cubic meter greater than or equal to 50 micrometers in minimum dimension	For organisms larger than 50 microns in minimum dimension: Discharge less than 10 per cubic meter of ballast water;
Discharge less than 10 viable organisms per milliliter less than 50 micrometers in minimum dimension and greater than or equal to 10 micrometers in minimum dimension	For organisms equal to or smaller than 50 microns and larger than 10 microns: Discharge less than 10 per milliliter (ml) of ballast water; and
Discharge of the indicator microbes shall not exceed the specified concentrations described in the following paragraph: Indicator microbes, as a human health standard, shall include: 1. Toxicogenic *Vibrio cholerae* (O1 and O139) with less than 1 colony forming unit (cfu) per 100 milliliters or less than 1 cfu per 1 gram (wet weight) zooplankton samples ; 2. *Escherichia coli* less than 250 cfu per 100 milliliters; 3. Intestinal Enterococci less than 100 cfu per 100 milliliters.	Indicator microorganisms must not exceed: (i) For Toxicogenic *Vibrio cholerae* (serotypes O1 and O139): A concentration of <1 colony forming unit (cfu) per 100 ml; (ii) For *Escherichia coli:* A concentration of <250 cfu per 100 ml; and (iii) For intestinal enterococci: A concentration of <100 cfu per 100 ml.

2.4. Applying Risk Assessment Principles to Ballast Water Management

The charge to the Panel focused primarily on whether BWMS could meet specific discharge standards, now or in the future. However, in its assessment the Panel found that combinations of practices and technologies should be considered as potentially more effective than reliance on BWMS technology (see section 6.5). Therefore this section sets the stage by exploring the use of risk assessment as a way to put strategies for treatment of ballast water into a probabilistic decision-making process. This process should be applied to the entire system of ballast water management and not to just one technique, device or practice. Each step of the process, from taking on ballast water at the port of origin to its discharge into the receiving port, depends upon the others. Risk assessment is a means of treating the entire ballast water management process in a holistic fashion and changes to each step can be evaluated within a defined risk-based process. A holistic approach to ballast water management includes the regulatory environment, the training of personnel, quality control, and environmental sampling. Risk assessment also provides the framework for risk management. One such framework is the Hazard Analysis and Critical Control Points

(HACCP). The HACCP process and its application to ballast water management are described in section 6.6.

2.4.1. Risk Assessment of Nonindigenous Species

The establishment of a nonindigenous species is the joint probability of how often species are introduced, the initial population size necessary to ensure reproduction, and the probability that organisms would find a suitable environment for propagation. This joint probability is low for any one species or specific shipping event. However, a large number of species can be transported via ship, and thousands of ships arrive at U.S. ports, creating a substantial probability that a nonindigenous species or a new pathogen will become established. Given that shipping is a major industrial activity that will continue far into the future, even a small probability for each ship and for each species will result in successful invasions. The goal of a ballast water management (BWM) program is to lower that probability, especially for particularly damaging species and pathogens. For a BWM program to be successful, the goals need to be specific and measurable, and the operational context needs to be understood. First, a model of the relationship between the number of organisms in ballast water and the likelihood of invasion or infection by a pathogen needs to be derived.

Probabilistic Approach to Deriving Risk Due to Nonindigenous Species

A foundation for the risk assessment for invasive species has been established (Drake 2004, 2005; Landis 2004). The process is density-dependent, with an increase in the density of organisms leading to an increase in population growth rate (Drake 2004). Modeling also has demonstrated the importance of a beachhead effect, where a population increases in a relatively isolated habitat patch before spreading to the remainder of the environment (Deines et al. 2005). Both Drake (2005) and Deines et al. (2005) recognize the importance of eliminating the organisms during the initial invasion event or destroying the beachhead in order to implement control.

Deines et al. (2005) used spatially explicit stochastic difference models and Drake and Lodge (2006) employed stochastic differential equations to model invasion events. In both studies the importance of understanding the stochastic aspects of colonization, the initial population size (propagule pressure) and density dependent effects on population growth were important in determining the probability of invasion. The actual dynamics of invasions were sensitive to initial conditions. The combination of stochastic and non-linear components results in a distribution of outcomes in both studies. This means that any relationship between propagule pressure and probability of invasion will be a distribution of outcomes. These foundations have been used to estimate risk in case studies (Drake et al. 2006; Colnar and Landis 2007). A risk assessment for managing ballast water invasion should have as its foundation the stochastic-non-linear nature of the invasion process.

Propagule Pressure and Invasion Relationships

It is possible to derive relationships between the number of organisms with an invasive potential (propagules) and the probability of an invasion over a specified amount of time. Such a relationship is described by the upper panel of Figure 2-1. It is assumed that the greater the number of propagules, the greater the probability that an invasive potential will be established. In this instance, it is assumed that the relationship is sigmoidal and has a threshold, but a number of curves are possible and the actual curve may be specific to the type of organism or environment. The solid line represents the central tendency of the relationship, with the dashed lines representing confidence intervals. Note that the confidence intervals include the possibility of a successful invasion even without propagule pressure from ballast water and also the likelihood of no invasion even with organisms escaping. After all, organisms can come from a variety of sources other than ballast water.

The lower panel of Figure 2-1 illustrates a process for setting targets for the number of organisms in ballast water. First a policy decision is made about an acceptable frequency of successful invasion over a specified amount of time. Existing information may be inadequate, so derivation of this frequency may require an assessment tailored to a specific habitat, species, endpoint, and location. Reading across the graph to where this rate intersects with the concentration-response curve gives the numbers of organisms corresponding to the low, expected and high values. Trade-offs can then be made on the likelihood of success in meeting the specified target and the costs of achieving the goal.

Although these graphs were drawn to express the relationship with one species of concern, similar plots may be derived for discharges with a large number of species. The greater the diversity of species and life stages, the greater the probability of an invasion by at least a single species.

2.4.2. Ballast Water Management Goals and the Decision Making, Risk Assessment Context

In order to evaluate the various types of BWMS, it is important to understand how they fit into a decision-making context. This means the management goal has to be clearly defined as in the graphical model (Figure 2-1). In addition, the effectiveness of ballast water treatment has to be evaluated within the context of a ship carrying cargo, human food and waste, and many organisms attached to the hull. The sea chest (a portion of the ship where seawater can be loaded or discharged) also can be a source of nonindigenous species. There is also the possibility of human error in the treatment process that may lead to the escape of organisms or the release of toxic materials. Each of these items will be covered in the paragraphs below.

The Goal: What Does "Zero" Mean?

What does "zero" discharge of nonindigenous species and other organisms mean as a goal, since such a value is essentially not measureable directly (see section 3)? The required sample volumes are enormous, there are refugia from treatment within the ballast water tanks, and the discharge is into an environment with multiple sources of invasive species. Operational definitions are very important and may prove more useful in making a decision about ballast water treatment options. For example, does "zero" mean that a discharge from a specific ship will contain no organisms that will colonize or infect a port environment for that one particular combination of disinfection treatment and vessel discharge? This is a very specific criterion but it is not necessarily protective. Furthermore, given the logistics needed to sample and enumerate organisms in a discharge, it will not be possible to meet this requirement for every discharge of every ship.

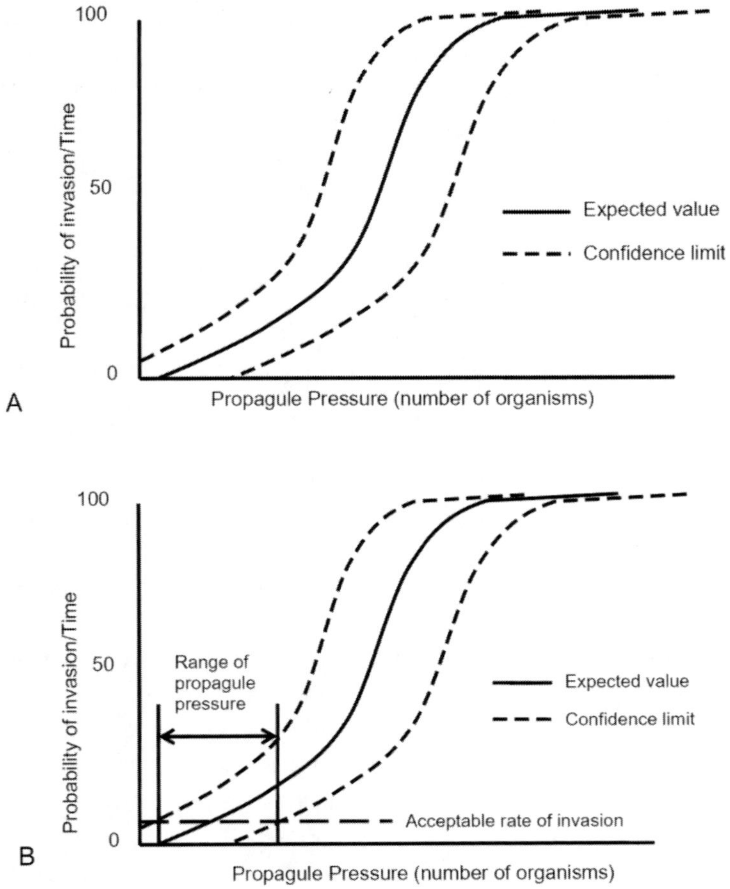

Figure 2-1. This conceptual diagram illustrates the relationship between propagule pressure (concentration) and the probability of an invasion by that species over time. The confidence intervals around the expected probability describe the uncertainty in the relationship (A). Such a curve will allow a quantitative determination of suitable goals for reducing potential invasions from nonindigenous species. Once a level of probability is agreed upon, the corresponding values of propagule pressures likely to produce the result can be obtained (B).

On the other hand, "zero" could mean the treatment technology or system will prevent introduction of a harmful invasive organism or disease to that port over a 10-year period. This is a very different criterion, a performance-based requirement that states the goal (no invasion or infection) over a specified time frame. Individual treatments on certain ships may fail, but an overall system would ensure that any colonizing organisms were quickly eradicated or that other methods would be employed to prevent their propagation. These two "zero" goals are very different and each puts onboard or land-based treatment options into specific and differing contexts. In order to rank the various technologies and treatment systems, therefore, the specific goals of the program need to be carefully defined.

There is also the question of specific goals for the protection of the port from nonindigenous species and pathogens. Are there specific requirements for each category of organism or is a combination approach to be attempted? Consider pathogenic organisms as an example.

In ballast water, a large proportion of the organisms likely are not pathogenic, but the human welfare implications may be higher for the pathogenic organisms than non-pathogenic organisms. Is the goal protection against human pathogens or those pathogens that may infect shellfish and fish populations, destroy important sea grass beds, or other segments of the ecological structure of the receiving port? Depending upon the specific policy goals, different propagule pressure-infection relationships may need to be considered.

Context of a Cargo or Tanker Ship and the Port Facility

As the specifications for the treatment process are made explicit, it is also important to understand the context of a ship and its port facility. Ballast water would be only one of the potential sources of nonindigenous species and pathogens brought by a vessel. Ships contain cargos of varied types, crew, food, human waste, and hull fouling organisms, each of which also could be sources of invasive species.

The port facility may also contain a variety of other vessels that may be sources of nonindigenous species and pathogens. Understanding the efficacy of the treatment program needs to be placed into this broader context.

Cargo may contain insects, fungi, seeds and spores that can be released to the environment as the cargo is unloaded or transported. Food can be another source of nonindigenous species, especially if living organisms are transported. Human waste can be a source of pathogens, but can be disposed of using appropriate facilities. Fouling of the hull of a ship can be a source of nonindigenous species or pathogens depending upon the origin of the ship, route and time of transit, and the effectiveness of the anti-fouling paint and the overall condition of the hull. The sea chest is a repository of organisms from across the travels of the vessel.

A confounding factor is that a number of other vessels will use the same port facilities, and all are potential sources of invasion. Fishing fleets and pleasure craft, for example, often take very long voyages and may transport nonindigenous species to a harbor. Also, these "other" vessels exist in regulatory environments different than those of cargo ships, barges, and tankers, regulations that may be less restrictive with respect to the transport of nonindigenous species. Although not directly affecting the infection potential of any single ship, these "other" vessels can confound determination of treatment effectiveness or identification of an invasive species' source.

So although there may be zero propagules in ballast water discharged at a facility, there will remain some probability of an invasion at the port. Hence there is a non-zero confidence interval in the example considered in Figure 2-1.

The risks due to invasion are not the only risks to be considered in BWM. It will be important to assess the potential impacts of decontamination and the effluent upon the environment. Does disinfection for pathogens increase the risk to the environment from the treatment?

The number of ships that use a port may also contribute to the trade-off. Decontamination activities that release an effluent with some residual toxicity may not pose an important risk to a facility that has a low volume, but may be important in a busier port. Some ports are very specialized. Port Valdez, Alaska specializes in the shipping of crude oil and some oil product. Cherry Point, Wash. is a port that currently receives crude from a limited number of sites to the refineries and bauxite for the smelter. Other facilities, such as New Orleans or Seattle-Tacoma, receive a variety of container ships and cargoes from across the world.

Shipboard emergencies, accidents, human error, and equipment failure should be considered in the risk analysis and decision-making process. At times, weather conditions or shipboard emergencies may preclude the operation of shipboard treatment facilities. Operator error or equipment failure may happen on shipboard or on-shore facilities just as it does in waste-treatment facilities.

However, in wastewater treatment facilities, strong programs of operator training and certification are established, unlike for shipboard BWMS. In parts of the U.S., hurricanes and northeasters can damage ships and on-shore equipment. No matter the weather, accidents and equipment failure will occur and will introduce nonindigenous species to a port facility. Maximizing reliability of the BWMS should be an important part of the risk analysis process.

BWM in an Overall Management Program

Large-scale establishment of species have occurred from what appear to be multiple invasions. Kolar and Lodge (2001, 2002) and Kolar (2004) describe examples for the Great Lakes in which populations of European fish have been established from multiple invasion events. European green crab was established in San Francisco in the late 1980s and the species has spread north along the west coast (Behrens and Hunt 2000). Invasions take time, often decades, are often due to multiple releases, and are difficult to control once established. A BWM strategy to decrease the rate of successful invasions should be part of an overall plan for the reduction of invasion events, monitoring, containment and eradication. Emphasis only on one aspect, the initial invasion event, is not likely to reduce the risk of successful invasions to an acceptable probability. The Hazard Analysis and Critical Control Points (HACCP) approach incorporates the context of the risk, potential points of control, and is very flexible in application. HACCP and its use for the management of invasive species are discussed in Section 6.6.

Summary

In summary, decisions about approaches to ballast water management can be viewed within a risk assessment framework. This framework should incorporate the following features:

- Recognition of the stochastic and non-linear nature of the invasion process;
- Clear definition of management goal needs; and
- Evaluation of the effectiveness of BWMS within the context of other sources of invasive species on the vessel, the treatment system, and the specific receiving habitat.

A ballast water management strategy to decrease the rate of successful invasions should be part of an overall plan to reduce invasion events, and their subsequent monitoring, containment, and eradication activities. Emphasis only on one aspect, the initial invasion event, is not likely to reduce the risk of invasions to an acceptable probability. Such management systems are addressed in Sections 6.5 and 6.6 of this report.

3. Statistics and Interpretation

3.1. Introduction

A consideration of statistical issues encountered in testing performance of BWMS is essential to the credible evaluation of the performance data – whether that evaluation is done by expert panels, testing facilities, or regulators.

This section presents key statistical considerations relevant to conditions under which BWMS performance is evaluated. A more detailed discussion of these issues is provided in Appendix C.

Testing conditions include the need to sample large volumes of water, particularly for the size class of organisms ≥ 50 μm in minimum dimension (nominally zooplankton, referred to hereafter as 'zooplankton-sized organisms') and to apply statistical methods that can quantitatively assess the confidence of test results obtained from counts of low numbers of organisms. These discussions pertain to both land-based and shipboard verification testing to determine conformity to a given performance standard in a type approval process; the same statistical theory applies to compliance testing by port state control officers for compliance or gross non-compliance (e.g., exceedance of a standard by orders of magnitude).

Credible testing requires the following process: Water must be collected and filtered to concentrate organisms into a manageable volume. The volume of ballast water carried by commercial ships ranges from a few thousand m^3 to more than a hundred thousand m^3. The volume that must be sampled following treatment is a small fraction relative to that total volume but, nonetheless, a large volume must be filtered to determine the number of live zooplankton-sized organisms.

This size class has the lowest concentration threshold – that is, organisms per m^3 vs. organisms per *ml* in the other two size classes – and represents the most challenging size class in terms of sampling to achieve statistical rigor. Hence, many of the examples in the following discussion will focus on zooplankton-sized organisms. The required sample volumes for these organisms are in the range of five to tens of m^3; the latter approximates the volume of a city bus.

In all size classes, subsamples of the concentrated volume are analyzed for viable (living) organisms, because all standards are based on the number of organisms surviving the treatment method. Once these counts are in hand, how reliably they portray conditions in the ballast water discharge must be determined. To accomplish this task, the live organism counts are analyzed using statistical methods to assess the uncertainty associated with the counts.

Assessing uncertainty in test results requires accounting for the spatial distribution of zooplankton-sized organisms in the sampled volume of water. Different probability distributions apply depending upon whether organisms are randomly distributed throughout a sample or are aggregated.

Therefore, this section illustrates how the use of appropriate probability distributions can characterize the level of reliability in taking the important inferential step from observing actual organism counts to determining whether a stated standard has been met.

3.2. Assessing Whether Ballast Water Standards Can be Met — The Statistics of Sampling

Without a well-defined, rigorous protocol based upon probability sampling, any standard will be difficult to assess and defend, and it will be impossible to compare the effectiveness of different BWMS. To outline what a sampling scheme might entail, and what sorts of information it would yield, it is necessary to investigate the probabilistic characteristics of plankton in ballast water.

Organisms can have one of two spatial characteristics: they can be randomly dispersed or clumped (aggregated) (see Lee et al. 2010). Because any sampling protocol is a function of the organisms' spatial distribution, it is critical to understand the distribution in the tank and discharge pipe and then sample accordingly. For randomly distributed organisms that are not abundant, the Poisson distribution can be used to estimate probabilities and conduct statistical power analyses (the probability that the sampling will find a vessel in or out of compliance when that is the case). Other hypervariable discrete alternatives to the Poisson distribution are available, such as the Poisson Log Normal and Poisson Inverse Gaussian distributions. The Panel chose to focus on the Poisson distribution because statisticians examining samples of treated ballast water have used the Poisson distribution, and the theoretical determination and empirical data collected thus far support its use. The Panel notes, however, that the negative binomial distribution is appropriate as the underlying statistical model for concentrations of organisms that are spatially aggregated.

3.2.1. The Poisson Distribution

Theoretical Considerations

The Poisson distribution has the property that its variance is equal to its mean, resulting in an increase in variability at higher densities. One way to assess whether the Poisson distribution is appropriate is to calculate the variance-to-mean ratio and compare it to 1.0. If a Poisson distribution is used, a single representative sample must be collected. To meet this requirement, the EPA Environmental Technology Verification (ETV) Generic Protocol for the Verification of Ballast Water Treatment Technology (U.S. EPA 2010) specifies that the sample be collected continuously over the entire discharge of the ballast tank and in an isokinetic manner. Assuming a given concentration, one can calculate the sample volume needed to guarantee a stated probability of finding at least a single planktonic organism (plankter) in that volume. An underlying assumption is that organisms are *randomly* distributed. Spatially aggregated populations present additional difficulties (see below), but if all organisms are counted in a continuously and isokinetically drawn representative sample, the issue of spatial distribution can be minimized or eliminated.

A major challenge of sampling at low organism concentrations is that many samples will have zero live organisms because the few live organisms present are missed. To improve the probability of detecting them, impractically large volumes must be sampled and excellent techniques must be used to enable detection (Figure 3-1). For example, when the water to be evaluated has a known concentration of one organism in 1 m^3, the probability of a 1 m^3 sample containing zero organisms is 36.8%; if the known concentration is 0.01 organism in 1 m^3 (equivalent to one organism in 100 m^3), the probability of obtaining a sample with zero organisms is ~99% (Lee et al. 2010; Appendix C). Furthermore, "If a small volume is used to evaluate whether the discharge meets a standard, the sample may contain zero detectable organisms, but the true concentration of organisms may be quite high.... *The general point is that more organisms may be released in ballast discharge using a stringent standard paired with a poor sampling protocol than a more lenient standard paired with a stringent sampling protocol*" (Lee et al. 2010, p.72, emphasis added).

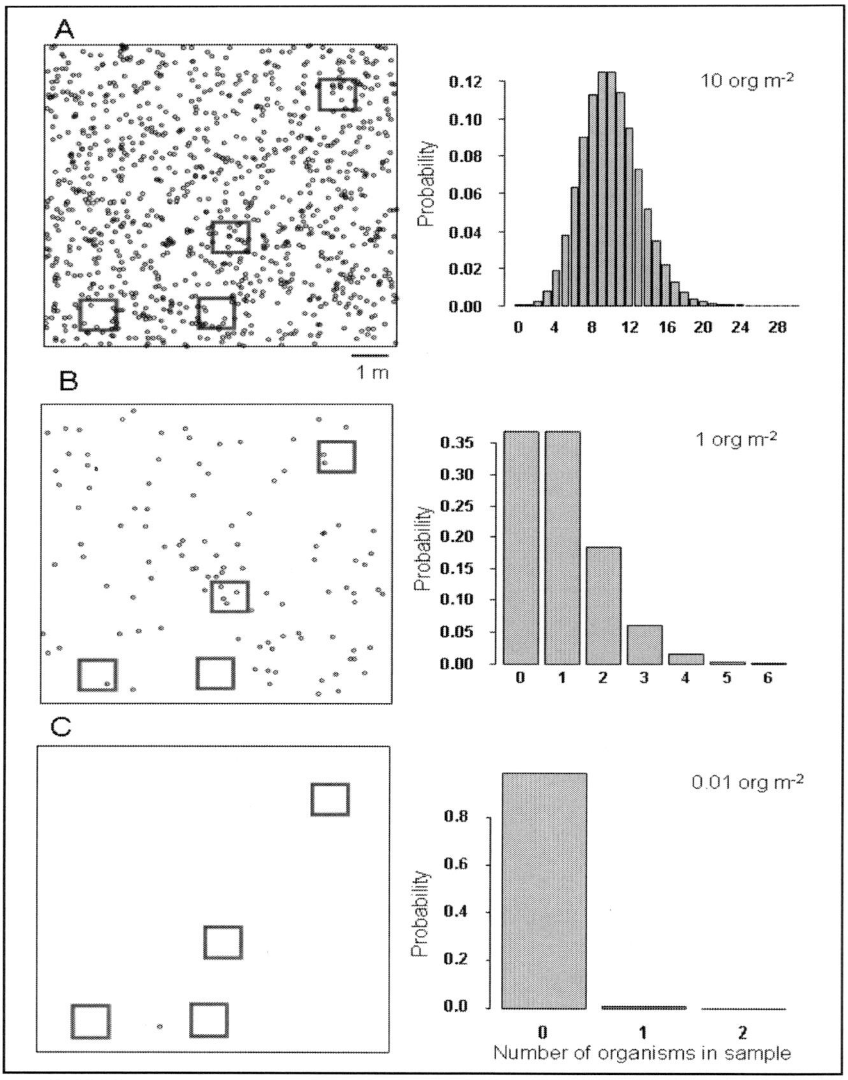

Source: Lee et al. 2010.

Figure 3-1. Illustration of the need to sample very large volumes to detect low concentrations of organisms present, assuming random distribution: Probability distributions for random samples of 1 m^2 for a randomly distributed population with 10 (A), 1 (B), or 0.01 (C) organisms m^{-2}. Red squares represent random samples. The data are displayed in terms of area with units of m^2, but the probabilities are the same for volumes. Plots on the right indicate the probability that a 1 m^2 sample will contain a given number of organisms. At low concentrations, the concentration of organisms likely will be estimated as 0 organisms m^{-2}, unless very large volumes are sampled.

Table 3-1. Sample volume of treated ballast water required relative to treatment standards for organisms ≥50 μm (nominally zooplankton), assuming that the desired level of precision of the estimated density is set at the 95% confidence interval of the Poisson distribution (= twice the observed mean and not greater than the standard limit). These are the required whole-water sample volumes that must be concentrated to 1 L as a function of N, the number of 20 1-mL subsamples analyzed

N=	1	3	5
Concentration (i.e., performance standard) (individuals m^{-3})	**Sample Volume Required (m^3)**		
0.01	60,000	20,000	12,000
0.1	6,000	2,000	1,200
1	600	200	120
10	60	20	12

Source: U.S. EPA 2010.

The ETV Protocol stipulates that biological samples (for all three size classes) should be continuously acquired on a time-averaged basis from a sampling port positioned in fully turbulent flow (U.S. EPA 2010), and are thus representative of the entire volume to be sampled. Organism abundance in BWMS testing can be statistically represented by the Poisson distribution and, therefore, the cumulative or total count is the key test statistic (Lemieux et al. 2008; Miller et al. 2011). A Chi-square distribution can also be used to approximate confidence intervals (CIs). However, experimental validation must be obtained to ensure that testing organizations can accomplish detection of live organisms with quantified uncertainty (see section 6.2.4 on viability).

The available methodologies for testing compliance with the IMO standards for zooplankton-sized organisms are at or near the analytic detection limits. For example, based on the Poisson distribution for a 95% CI from the Chi-square distribution, 30 m^3 (30,000 L) must be sampled in order to find and count < 10 organisms m^{-3} with the desired level of precision (U.S. EPA 2010; Appendix C).

The ETV Protocol provides examples of the sample size needed to provide the level of precision needed to achieve a 95% upper confidence limit that is no more than twice the observed mean and does not exceed the targeted concentration (Tables 3-1 and C-1, Appendix C; U.S. EPA 2010). If the volume of subsample that is analyzed is increased, then validation experiments should be conducted to ensure that counting accuracy is acceptably high. The Poisson distribution assumption still applies to organisms in the next smaller class (here, referred to as "protist-sized" organisms, or organisms ≥10 μm and < 50 μm in minimum dimension), and the ETV Protocol provides examples with a more stringent level of precision than is used for the larger size class (Table C-1, Appendix C; U.S. EPA 2010). At present, confirmation of the Phase 1 standard (< 10 protist-sized organisms mL^{-1}) represents the practical limit that can currently be achieved by testing facilities in the U.S. (e.g., MERC 2009a, 2010a, 2010b; Great Ships Initiative 2010). In addition, determining viability of protist-sized organisms remains problematic because many organisms do not move during time scales over which they are observed (as do many zooplankton; see section 6.2.4 for a discussion of viability determination).

It is expected that the statistics governing the smallest size classes—the < 10 μm protists proposed below, and the indicator and pathogenic bacteria *(Escherichia coli, Enterococci* spp., and *Vibrio cholerae)*—will be similar to the two size classes discussed here. That is, treated samples will be well- mixed and will have sparse populations collected from representative samples of ballast water; thus, the Poisson distribution would be applicable.

Laboratory Experiments

Laboratory experiments with protist cultures support use of the Poisson distribution (Nelson et al. 2009; Steinberg et al., accepted with revisions; Appendix C). Based on results of laboratory experiments with two sizes of standardized microbeads at different densities as imperfect proxies of protist- and zooplankton-sized particles, Lemieux et al. (2008) recommend that samples for analysis of protist-sized organisms be concentrated by at least a factor of five, and that at least four replicate counting chambers should be analyzed for acceptable accuracy and precision (see Appendix C for details). Furthermore, Lemieux et al. (2008) determined that the zooplankton size class requires a sample size of greater than 6 m^3 concentrated to 0.5 L and analysis of at least 450 1-mL aliquots. Because these higher concentration factors are likely unrealistic, Lemieux et al. (2008) suggest that larger sample sizes and improved analytical methods be used.

When concentrations are close to the performance standard, a single sample may require too large a volume of water to be logistically feasible. In that case, complete, continuous time-integrated sampling (with the entire volume analyzed) and combining samples across multiple trials can improve resolution while maintaining statistical validity. To that end, Miller et al. (2011) applied statistical modeling (based on the Poisson distribution) to a range of sample volumes and plankton concentrations (see Appendix C for details). They calculated the statistical power of various sample volume and zooplankton concentration combinations to differentiate various zooplankton concentrations from the proposed standard of < 10 live organisms m^{-3}. They concluded that three trials of time-integrated sampling of 7 m^3 from a ship's ballast water discharge theoretically can result in 80% or higher probability of detecting noncompliant discharge concentrations of 12 vs. 10 live organisms m^{-3}. Thus, pooling volumes from separate trials will allow lower concentrations to be differentiated from the performance standard, although the practicability and economic costs of doing so have not been evaluated.

It is important to note, however, the practical limits of increasing statistical sample sizes that may already tax the capabilities of well-engineered land-based ballast water test facilities used in verification testing. Shipboard testing in the U.S. has been done on a pilot scale to date (i.e., the USCG Shipboard Testing Evaluation Program, STEP), but pooling volumes from multiple trials might also be problematic on vessels used for shipboard verification testing and compliance testing. According to Table 3-1, to meet a standard 10 times more stringent than D-2/ Phase 1 would require anywhere from 120-600 m^3 of whole-water sample volumes, which is impracticable; test facilities in the U.S. typically analyze ~5 m^3 of water per test (e.g., MERC 2009a, 2010a, 2010b; Great Ships Initiative 2010).

Additional Challenges of Sampling Large Volumes

As outlined in Lee et al. (2010), the detection of viable organisms at very low concentrations is a major practical and statistical challenge, partly because of the inherent stochasticity of sampling. Due to random chance, the number of organisms in multiple samples taken from the same population will vary. In addition, very large volumes of water must be sampled in order to accurately estimate the organism densities. Three other considerations merit attention.

First, statistical approaches rest upon the premise that the samples realistically represent the actual concentrations of organisms discharged. This premise is based upon two assumptions: all organisms are detected in the analyzed volume (no human or equipment errors), and organisms are randomly distributed in both ballast tanks and discharge water. Neither assumption will be true all of the time. Human and equipment errors will occur, and organisms are typically "patchy" or non-random within the water column of a tank or the stream of a large-volume discharge (Murphy et al. 2002; U.S. EPA 2010). If appropriate quality control and assurance procedures were used in collecting the data, then ideally human error and equipment malfunction would have been accounted for. The degree of randomness can be determined by calculating the variance-to-mean ratio from multiple samples and comparing the resulting ratios to 1.0. Additionally, analysis of a time-averaged sample taken continuously in an isokinetic manner renders this assumption moot.

Second, the logistics of managing large sampling containers, sample transport costs (since samples usually are not processed aboard ship), analytical supplies, and personnel time would make it impractical to process all of the volume of, for example, even one 100 m^3 sample, much less multiple samples, especially in type approval of BWMS when multiple successful tests are required. Lee et al. (2010) calculated the probability of finding one or more organisms in a sample for a series of organism concentrations and sample volumes (Table C-2, Appendix C). These calculations show that 100 L of ballast must be sampled to have a > 99% probability of detecting at least 1 zooplankton-sized organism when the true concentration is 100 organisms per m^3. When small sample volumes are collected, the probability of detecting an organism is low even at relatively high organism concentrations; for example, organisms will be detected in fewer than 10 percent of subsamples if a 1-L sample is taken and the "true" concentration is 100 organisms m^{-3}. Lee et al. (2010) then estimated the upper possible concentration (UPC, upper 95% CI) of organisms actually present in ballast water from the number of organisms in a sample volume based on the Poisson distribution. Zero organisms detected in a 1-m^3 sample could correspond to a true concentration of organisms in the ballast tank of up to ~3.7 organisms m^{-3}; if the sample volume is only 1 L, zero organisms detected could correspond to a true concentration of ~3,700 organisms m^{-3} (Table C-3, Appendix C).

Third, in the above analyses, the true concentrations are known. The goal in sampling unknown concentrations is to accurately assess whether a given BWMS treats water with true organism concentrations that meet a given performance standard. Inherent stochasticity of sampling may result in an indeterminate category, as well, and the probability of obtaining an indeterminate evaluation increases with decreasing sample volume and increasing stringency of the ballast water standard (Figure C-1, Appendix C). For example, it would be necessary to sample ~0.4 m^3 of ballast water to determine whether the D-2/Phase 1 standard of < 10 zooplankton-sized organisms m^{-3} was met if fewer than approximately 10 organisms were observed in the sample (Figure C-1B, Appendix C).

3.2.2. Spatially Aggregated Populations – Negative Binomial Distributions

This section illustrates how difficult statistical analyses can become when working with spatially aggregated populations and emphasizes the gains made from doing a complete count of a representative sample that has been continuously and isokinetically taken.

If organisms are aggregated rather than randomly distributed in a ballast tank, a different statistical approach is required. For aggregated populations, the variance exceeds the mean (negative binomial distribution, $\sigma^2 > \mu$); thus, as the variance increases, the number of organisms in a random sample is increasingly unpredictable. Lee et al. (2010) recommend use of the negative binomial distribution to model aggregated populations.

Because it is more difficult to accurately estimate the true concentration, more intensive sampling is required. For a randomly distributed population with a true concentration of 1 organism m^{-3}, ~37% of the subsamples from a 1 m^3 sample of treated ballast water would contain zero organisms; for an aggregated population with a dispersion parameter of 0.1, ~79% of the subsamples would contain zero organisms (Figure C-2, Appendix C; Lee et al. 2010). As aggregation increases, the probability of samples containing either zero organisms or large numbers of organisms relative to the true concentration also increases (Figure C-3, Appendix C). Thus, large numbers of subsamples from large sample volumes must be taken to account for aggregated populations; otherwise, there will be a high probability that the concentration estimates from sample analyses will be either much lower or much higher than the true concentration.

Determination of whether a population is aggregated is complicated, since it is the scale of the aggregation pattern relative to the size of the sampling unit that controls the estimate of aggregation (Figure C-3, Appendix C). Lee et al. (2010) recommend the Taylor power law (Taylor 1961) as an alternative to the negative binomial, because it can accommodate a wider range of aggregated distributions than the negative binomial.

Overall, the degree of aggregation represents challenges in sampling sufficiently large volumes of ballast water to determine whether a given BWMS passes or fails to meet standards more stringent than the present IMO guidelines, even if the true concentrations of organisms are 10 to 1000 times higher than the performance standard. This remains a problem in quantifying many protist-sized organisms, but becomes less of a problem with very small organisms such as bacteria, which have a tendency to clump but are effectively counted as colonies and not individuals. However, laboratory experiments provide data supporting use of the Poisson distribution to analyze ballast water samples (Lemieux et al. 2008; Nelson et al. 2009; see Appendix C).

3.3. Interactive Effects

A final consideration regarding statistical analysis is the potential for covariance, or interactive effects among environmental conditions – for example, a treatment system may perform well under high- temperature or high-biomass conditions, but not both (Ruiz et al. 2006). To address this problem, covariate measurements should be carefully addressed in experiments, and treatment evaluations should consider the potential for interactions and target tests of especially challenging combinations.

3.4. Certainty of Results

As with all statements that are based upon statistical sampling, there is always a stated non-zero error probability associated with a particular statistical conclusion. Thus, without a complete census of a ship's entire volume of ballast water, one can never claim to be 100 percent certain that the concentration of live zooplankton-sized organisms is below a discharge standard. Available methodologies to test D-2/ Phase 1 compliance are presently at or near analytic detection limits for the two largest organism size classes. The D-2/ Phase 1 performance standards are measurable at present based on land-based and shipboard testing approaches, however, new or improved methodologies will be required to increase detection limits.

From the examples above, statistical theory shows it is theoretically possible to detect adherence to a very low discharge standard (e.g., 1000x more stringent than D-2/ Phase 1). Measuring to a 1000x more stringent standard, however, is impracticable at the present time because of the logistics of collecting, reducing, and counting organisms in all size classes within the volumes of water required. Detecting achievement of a standard ten times more stringent may be possible (although consensus has not been reached on this), but it seems unlikely for the reasons mentioned above that detecting achievement of a 100x more stringent standard is possible.

3.5. Conclusion

- Rigorous statistical sampling protocols (that may include consideration of the spatial distribution of plankton in ballast water) and subsequent statistical analysis are required to assess whether a BWMS meets desired performance standards.
- Detecting organisms in low abundance is a difficult problem that requires sampling of very large volumes of water, especially for the zooplankton-sized organisms ($\geq$ 50 μm).
- The initial sample volumes needed are a function of the degree to which the sample volumes are concentrated, the performance standard, and the desired level of confidence (e.g., 95%, which is used most often in ecological investigations).
- The Poisson distribution is recommended as the model for statistical analysis of treated water samples.
- Available methodologies to test D-2/ Phase 1 compliance are presently at or near analytic detection limits for the two largest organism size classes. New or improved methodologies will be needed to increase detection limits.
- The D-2/ Phase 1 performance standards are measureable at present. Because of the logistics of collecting, reducing, and counting organisms in all size classes within the volumes of water required to achieve a standard 1000 times more stringent than the D-2/ Phase 1 performance standard, measuring adherence to a 1000x more stringent standard may be impracticable. Measuring adherence to a standard that is 10x more stringent may be possible if a continuously isokinetically taken representative sample is used. It seems unlikely, for reasons mentioned above, that a 100x more stringent standard could be measured at present.

- Statistical conclusions at a stated confidence level always have an associated error probability; thus, "100 percent certainty" is not statistically possible without a complete census of a ship's entire ballast water contents.

4. Performance of Shipboard Systems with Available Effluent Testing Data

4.1. EPA's Charge Questions

This section responds to Charge Questions 1 and 2, which ask the Panel to assess the documented performance of existing BWMS in terms of quality of discharged ballast water, and to assess the likely future performance of BWMS based on their design and treatment processes.

4.2. Assessment Process

A subgroup of the Panel led the assessment of BWMS technologies. The subgroup considered only the information compiled by EPA through solicitation of various Maritime Administrations that have granted Type Approval certifications, direct communication with developers and manufacturers of BWMS, and searches for publically available sources (such as journal or conference publications and third-party reports available through the Internet). This information (listed in Appendix A) included data packages, reports, publications, certification documents, and other available information on the performance of BWMS.

Three subgroup members independently examined in detail all data packages, with two other members providing review oversight and quality control. The type, amount, and quality of material in the data packages varied—some contained only a type approval certificate, while others included land-based and shipboard testing methods and data, documentation of G9 approval, a type approval certificate, and press releases describing the sale of systems for use on commercial vessels. The Panel notes that BWMS are still evolving with an ever-growing number of manufacturers developing systems. Thus, this analysis represents a snapshot in time. No new data or information was considered beyond packages submitted to the SAB by December 1, 2010.

4.3. Assessing the Reliability of Existing Data

The three primary reviewers independently scored each package as having "reliable" or "unreliable" data. To earn a "reliable" rating, the data package had to include, at a minimum, methods and results from land-based or shipboard testing. A BWMS holding a certificate of type approval without supporting testing data was scored as having "unreliable" data because it was impossible to determine the validity of the testing procedures and, therefore, the validity of the data. If a BWMS's data package included one or more test reports, the data package was examined according to the following criteria:

- The operational type of system (e.g., deoxygenation + cavitation) was determined to be generally appropriate for shipboard use (e.g., can it meet required flow capacities, size, and power requirements).
- The technical literature supported the fundamental use of the technologies used (e.g., is it well documented that using the approach will safely and effectively remove, kill, or inactivate aquatic organisms).
- Laboratory testing was conducted with "reasonable and appropriate methods" (i.e., methods commonly used in aquatic studies or alternative methods that appear rigorous and equivalent to a standard, common approach).
- Land-based testing was conducted with reasonable and appropriate methods; sample number and size were appropriate; sample collection and handling was appropriate and documented; analytical facilities were adequate; IMO or ETV (v. 5.1) challenge conditions were met; if necessary, toxicological studies were conducted and demonstrated environmental safety; a QA/QC policy was in place and followed; and ultimately, land-based testing produced credible results.
- Shipboard testing was conducted with the same considerations as land-based testing (described above) and produced credible results.
- If an active substance was included, the BWMS had credible toxicity and chemistry data and G9 Basic approval or G9 Final Approval (which requires Basic approval).
- The BWMS had a type approval certificate.
- The BWMS was in operational use (i.e., not used only during shipboard type approval testing) on one or more active vessels. A BWMS not yet having operational systems onboard vessels was not automatically categorized as having "unreliable" data, but this information was useful.

It is important to note that if the data packages were deemed "reliable, it was assumed that all protocols and methods were followed exactly as described. For data packages that included clear QA/QC procedures, there was a higher level of certainty that this was the case. In the absence of QA/QC documentation, which was the case for most data packages, the level of rigor in following the protocols and methods described was unknowable.

4.4. Assessing the Ability of BWMS to Meet Discharge Standards

For BWMS with reliable data, the system's ability to meet four discharge standards—IMO D-2/ USCG Phase 1 and 10x, 100x, and 1000x more stringent than IMO D-2/USCG Phase 1—was determined, again independently, by the three primary reviewers.

- **10x** was evaluated based on BWMS ability to reduce concentrations of living organisms: (a) ≥50 μm in minimum dimension to below 1 per m^3, (b) ≥10 to <50 μm in minimum dimension to below 1 per ml, and (c) a decrease in total bacteria.
- **100x** was evaluated based on BWMS ability to reduce concentrations of living organisms: (a) ≥50 μm in minimum dimension to below 1 per 10 m^3, (b) ≥10 to <50 μm in minimum dimension to below 1 per 10 ml, and (c) and a significant reduction in total bacteria.

- **1000x** was considered the equivalent of the **USCG Phase 2** standard, including: (a) ≥50 μm in minimum dimension to below 1 per 100 m^3, (b) ≥10 to <50 μm in minimum dimension to below 1 per 100 ml, (c) total bacteria below 10 per ml, (d) total viruses below 100 per ml, and below levels listed for indicator microbes (see Table 2-1).

The following scores and interpretations were assigned:

A - Demonstrated to meet this standard in accordance with the approach suggested in the IMO G8 guidelines (and G9 guidelines, if the BWMS employs an active substance).
B - Likely to meet this standard if the more detailed ETV Protocol (and corresponding sample volumes) were to be used.
C - May have the potential to meet this standard with reasonable/feasible modifications to the existing BWMS.
D - Unlikely, or not possible, to meet this standard, even with reasonable/feasible modifications to the existing BWMS.

To date, all BWMS adjudged to have reliable data have been tested in accordance with the G8 guidelines, which provide only general recommendations for how to evaluate performance with respect to the D-2 standards. In late 2010, EPA's Environmental Technology Verification (ETV) Program released the Protocol for the Verification of Ballast Water Treatment Technologies (Version 5.1, EPA 2010). Although no BWMS has yet been tested under the ETV Protocol, this protocol provides much more detailed instructions for how to conduct BWMS tests that are scientifically rigorous and statistically sound. In particular, the ETV Protocol has significantly improved sampling procedures. The IMO G8 guidelines suggest collecting replicate samples with volumes of at least 1 m^3 for the size class of organisms ≥ 50 μm in minimum dimension (nominally zooplankton). ETV, and others have demonstrated that a time-integrated sampling approach with larger sample volumes will increase statistical confidence regarding whether zooplankton in sparse populations meet or exceed the IMO D-2/Phase 1 standard (Miller et al. 2011; Lee et al. 2010; section 3, above). As such, although D-2 and Phase 1 standards are essentially the same, some BWMS were given a score of 'A' if the data showed they met the D-2 standard by following the G8 guidance, and received a 'B' for Phase 1 if the number of living organisms was consistently low and it seemed very likely the BWMS would still meet the standard if ETV Protocols (including larger, integrated samples) were used. Regarding the discharge standard 10x more stringent than the IMO D-2/ Phase 1, the criterion used was whether the number of living organisms in all size classes was consistently low following testing (below the detection limit, often reported as zero, or not more than twice the standard). If so, the BWMS was given a 'C', indicating it had the potential to meet the standard. However, as described in the response to charge question 4 (Section 6), current testing methods do not provide the resolution required to conclude that 10x standards can be met. For the most stringent standards, 100x and 1000x more stringent than IMO D-2/ Phase 1, if any living organisms in any size class were found following treatment, the BWMS earned a 'D'. This score indicates that it is extremely unlikely (or perhaps impossible) that the BWMS could meet a stricter standard, again because the detection limit of the test methods used provide resolution to IMO D-2/

Phase 1, at best. For example, if one viable zooplankter was found in testing using volumes of 1 m^3, the BWMS would be required to reduce the number of viable zooplankters to less than one in 10 m^3 or 100 m^3 to meet the 100x and 1000x standards, respectively. After each subgroup member completed his or her individual, independent assessments, they discussed their scores collectively. All scores from the three primary reviewers were found to be identical and in complete agreement with general assessments by the two subgroup oversight members, as well as other members of the entire Panel. These consensus findings were used to create Table 4-1. Rather than present the scores from individual, commercial BWMS units or models, the Panel categorized the technologies by operation type (e.g., filtration + UV). The operation types were chosen from recently published, third-party data reports (Albert et al. 2010; CSLC 2010; Lloyd's Register 2010) in order to encompass all currently available operation types and to provide a standardized terminology. Thus, while the data packages from individual BWMS were initially examined and scored, the results were aggregated to represent a top-order status of the field. For a given operation type, if reliable data were available for more than one commercial BWMS, the scores given to the operation type were the highest scores of any of the individual BWMS. In this manner, Table 4-1 represents the greatest potential for each of the operational categories of technologies to meet various discharge standards.

4.5. Assessment Results

The results of this assessment are presented in Table 4-1 and interpretations of the findings are provided below. For this assessment, 51 individual BWMS were identified from prior reports (Albert et al. 2010; CSLC 2010; Lloyds Register 2010) to show the breadth and diversity of treatment approaches. However, it is important to note that of the 51 BWMS listed, a large proportion are at early conceptual/development stages (only approximately 15 to 20 have been tested onboard an active vessel) and a few have recently been discounted because of logistic or performance challenges. The Panel received information packages on 15 individual BWMS, but just nine BWMS were considered to have reliable data for an assessment of performance.

Table 4-1. Performance of Ballast Water Management Systems

Type or Category of BWMS	# BWMS	# Type Approval Cert	# Available/ Reliable Data	D-2	P-1	10x	100x	1000x
Deoxygenation	2	0	0					
Deoxygenation+cavitation	1	1	1	A	B	C	D	D
Deoxygenation+bioactive agent	1	0	0					
Electrochlorination	2	1	0					
Electric pulse	1	0	0					
Filtration	1	0	0					

Table 4-1. (Continued)

Type or Category of BWMS	# BWMS	# Type Approval Cert	# Available/ Reliable Data	D-2	P-1	10x	100x	1000x
Filtration+chlorine	2	0	0					
Filtration+chlorine dioxide	1	0	1	A	B	C	D	D
Filtration+coagulation	1	1	0					
Filtration+UV	10	3	3	A	B	C	D	D
Filtration+UV+Ti O_2	1	1	1	A	B	C	D	D
Filtration+ultrasound	1	0	0					
Filtration+ozone+ultrasound	1	0	0					
Filtration+UV+ozone	1	0	0					
Filtration+electrochlorination	5	1	2	A	B	C	D	D
Filtration+UV+ozone+electrochlorination	1	0	0					
Filtration+electrochlorination+ advanced oxidation	1	0	0					
Filtration+cavitation+electrochlorination	1	0	0					
Filtration+-electrochlorination+ultrasound	1	0	0					
Filtration+cavitation+ozone+ electrochlorination	1	1	0					
Filtration+plasma+UV	1	0	0					
Filtration+cavitation+nitrogen+ electrochlorination	1	1	0					
Filtration+hydrocyclone+electrochlorination	1	0	0					
Heat	1	0	0					
Hydrocyclone+filtration+peracetic acid **	1	1	1					
Hydrocyclone+electrochlorination	2	0	0					
Hydrodynamic shear+cavitation+ ozone	1	0	0					
Hydrocyclone+filtration+UV	1	0	0					
Menadione	1	0	0					
Mexel	1	0	0					
Ozone	1	1	0					
Ozone+cavitation	1	0	0					
Shear+cavitation+ozone	1	0	0					
Shear+cavitation+peracetic acid	1	0	0					
Totals	**51**	**12**	**9**					

Green rows designate the types of BWMS that had reliable data and whose performance was evaluated against various discharge standards.

Based on one or more reliable data sets, the type of BWMS:

(A) demonstrated to meet this standard in accordance with G8/G9

(B) likely to meet this standard if more detailed ETV Protocols were used

(C) potential to meet this standard with reasonable/feasible modifications

(D) unlikely, or not possible, to meet this standard

** Not scored because the one manufacturer has withdrawn this BWMS from market

4.6. Response to Charge Question 1

The analysis described above formed the basis of the Panel's responses to charge Question 1a, 1 b, and 1 c; each of these sub-questions addresses different aspects of treatment capabilities for shipboard systems. These questions and our responses are summarized below.

Question 1a: For the shipboard systems with available test data, which types or categories have been evaluated with sufficient rigor to permit a credible assessment of performance capabilities in terms of effluent concentrations achieved (living organisms/unit of ballast water discharged or other metric)?

Conclusion 1a: Five types or categories of BWMS have been evaluated with sufficient rigor to permit a credible assessment of performance capabilities. These technology combinations are:

- Deoxygenation + cavitation
- Filtration + chlorine dioxide
- Filtration + UV
- Filtration + UV + Ti O^2
- Filtration + electrochlorination

Question1b: For those types or categories of systems identified in 1a, what are the discharge standards that the available data credibly demonstrate can be reliably achieved? Furthermore, do data indicate that certain systems (as tested) will not be able to reliably reach any or all of the discharge standards?

The five types of BWMS listed above have been demonstrated to meet the IMO D-2 discharge standard, when tested under the IMO G8 guidelines, and will likely meet USCG Phase 1 standards, if tested under the more detailed ETV Protocol. This level of treatment efficacy results in a 10,000x reduction in numbers of living organisms ≥ 50 μm in minimum dimension (under land-based testing guidelines). This represents an important achievement in the ability of these systems to effectively, reliably, and dramatically remove live organisms from ballast water under the challenging conditions found on active vessels.

The detection limits for currently available test methods and approaches prevent a complete statistical assessment of whether BWMS can exceed the IMO D-2/ Phase 1 standard. However, one way to predict the ability of a BWMS to meet 10x, 100x, or 1000x standards, is to consider the frequency with which any live organisms are detected during testing. This approach provides insight into BWMS consistency/reliability and its lower performance limits. Three frequency categories were defined using available data:

1. BWMS always produced "zero" or "non-detectable": The system is consistently exceeding current detection limits and thus the IMO D-2/ Phase 1 standards (as described above). However, if results for all test trials, for all categories of organisms, and for all samples from a specific BWMS reported "zero" or "non-

detectable," there is no way to determine if the system is performing just below the IMO D-2/ Phase 1 standards or if it is approaching 10x, 100x or 1000x.

2. BWMS produced "zero" or "non-detectable" most of the time, with only one or a very few readings above the detection limit: The system appears to be operating near but below the IMO D-2/ Phase 1 standards. It is also possible that the occasional or rare "detects" were a result of BWMS malfunction or an error in sample collection, handling or analysis.
3. BWMS produced results in the detection limit most of the time, with only one or very few "zero" or "non-detectable". The efficacy level of the system is clearly only at, or just below, the IMO D-2/ Phase 1 standards.

Not one of the BWMS examined could be categorized in group 1 (i.e., consistently scored "zero" or "non-detectable"). Instead, BWMS were roughly split between frequency categories 2 and 3. For all BWMS, live organisms in the ≥ 50 μm and/or ≥ 10 to < 50 μm size classes were detected in at least two independent test trials and in general, live organisms ≥ 10 μm were detected in 20% to 80% of test trials. It is also important to note that when total bacteria was quantified during the testing of BWMS, treatment did not reduce levels to that required in the 1000x discharge standard (10/ml). In fact, it was not uncommon to find an increase in total bacteria after treatment. Therefore, even if testing detection limits are improved, the lower performance limits of current BWMS are not expected to change.

Conclusion 1b: The Panel concludes that five types of BWMS are currently able to reach IMO D-2/ Phase 1 standards. These same five types may be able to reach 10x IMO D-2/ Phase 1 standards for the ≥50 μm and ≥ 10 to < 50 μm size classes in the near future, if both treatment performance and testing approaches improve (see Section 5). Finally, no current BWMS types can meet a 100x or 1000x discharge standard.

Question 1c: For those systems identified above, if any of the system tests detected "no living organisms" in any or all of their replicates, is it reasonable to assume the systems are able to reliably meet or closely approach a "no living organism" standard or other standards identified in Table 4.1 of the EPA White Paper (June 2010), based on their engineering design and treatment processes?

To address this question, the phrase "no living organisms" was considered in two distinct ways: first, in a literal sense to mean the sterilization of ballast water and second, from a scientific perspective to mean results below method detection limits.

Based on the test data provided for several BWMS, it is clear that numbers of live organisms in discharged ballast water are reduced dramatically relative to intake water and corresponding control water. Five distinct BWMS types have been demonstrated to meet the IMO D-2 standard and appear very likely to meet the USCG Phase 1standard (which demands, at minimum, a 4-log reduction from initial concentrations for the largest organism size class), not only in land-based testing but also under the physically challenging conditions presented on active merchant vessels during shipboard testing. However, even high levels of organism removal do not achieve sterilization or the complete removal of all living organisms. The identification of just one live organism would indicate non-sterile conditions, and all systems evaluated had at least one living organism in at least one treatment sample (and often more, as described above). Unfortunately, in some cases, this low number of live organisms might result from contaminated scientific sampling gear (nets, glassware, etc.) or human counting error.

Alternatively, it is possible to establish specific detection limits (e.g., 100, 10, 1.0, 0.1, live organisms per m^3 or ml) associated with the methods used to collect the current performance data available and thus to conclude that, if numbers of live organisms are below those detection limits, they are statistically indistinguishable from zero or no living organisms. Efforts have been made to calculate the probabilities of meeting such specified detection limits under certain assumptions, such as whether the organisms are randomly dispersed in space or spatially aggregated (see Lee et al. 2010 and Section 3 for details and examples). Not surprisingly, increased statistical power comes not only from increased sample size, but also from the difference between the mean established by regulation and the measured mean from a sample—which indicates the degree of compliance (or noncompliance). (See Section 3 for a more detailed discussion of sampling statistics and detection limits.)

Conclusion 1c: It is not reasonable to assume that BWMS are able to reliably meet or closely approach a "no living organism" standard. Available data demonstrate that current ballast water management systems do not achieve sterilization or the complete removal of all living organisms.

4.7. Response to Charge Question 2

Question 2: Based on engineering design and treatment processes used, and shipboard conditions/constraints, what types of ballast water treatment systems can reasonably be expected to reliably achieve any of the standards, and by what dates? Based on engineering design and treatment processes used, are there types or categories of systems that conceptually would have difficulty meeting any or all of the discharge standards?

A variety of BWMS types are being used to manage ballast water (Table 4-1). The data indicate that several types or categories are proving reliable and effective, and Table 4-1 lists five types that have been demonstrated to meet the IMO D-2/ Phase 1 standard. These five BWMS also appear to be mature technologies, with multiple active vessel installations, and are commercially available. Interestingly, four of the five treatment approaches include a filtration step, although the inclusion of filtration does not necessarily ensure that the BWMS will meet discharge standards. A large majority of BWMS also appear to be adapted from technologies long applied to water treatment.

Given the data available, it is reasonable to assume that these same five systems have the potential to meet a 10x IMO D-2/ Phase 1 standard in the near future (see Section 5). As noted above, the Panel makes this prediction based upon available data that show viable organisms sampled as low (usually, below detection limits) but improvements to test methods/approaches will be required to demonstrate conclusively that improved BWMS meet standards beyond IMO D-2/ Phase 1. Given the data available, it is highly unlikely that any of the systems listed in Table 4-1 could provide organism removal to the level of 100x or 1000x the standard because all systems showed at least one observation of a living organism within the sample volumes as specified in IMO D-2 guidelines, thus clearly exceeding these more stringent standards. No BWMS reported zero living organisms in all samples analyzed following treatment. In fact, most results showed an increase in total bacteria abundances after treatment, far exceeding discharge levels proposed in the USCG Phase 2 standards. Ultimately, different technologies or treatment approaches, and sampling strategies will be needed to achieve these higher levels of removal. At this time, it is not possible to comment on the likelihood that the other treatment types listed will, or will not, be able to meet either the IMO D-2/Phase 1 or more stringent standards. All the BWMS types listed in Table 4-1 have likely shown some potential for reducing the number of ballast water organisms, but for most the data available for examination were deemed either to be absent or unreliable. As such, it is not possible to predict the eventual performance of these BWMS.

Conclusion 2: Five types or categories of ballast water management systems can currently meet the IMO D-2 discharge standard and appear to meet USCG Phase I standard: Deoxygenation + cavitation; Filtration + chlorine dioxide; Filtration + UV; Filtration + UV + TiO^2; and Filtration + electrochlorination. It is possible that the same five types could meet 10x IMO D-2/Phase 1 sometime in the near future if both treatment performance and testing methods and approaches (e.g., detection limits) improve. Nearly all of the 51 treatment types or categories evaluated are based on reasonable engineering designs and treatment processes, and most are adapted from longstanding industrial water treatment approaches. However, the lack of detailed information on the great majority of BWMS prevents an assessment of limitations in meeting any or all discharge standards.

4.8. Environmental Effects and Vessel Applications: Additional Constraints and Considerations that Influence BWMS Performance

BWMS are still evolving with an ever-growing number of manufacturers developing systems. Although several BWMS have received type approval certification, and appear to safely (e.g., received final G9 approval) and effectively meet IMO D-2/Phase 1 discharge standards (Table 4-1), there are several factors to consider beyond mechanical and biological efficacy. Perhaps the four most important considerations for the broad applicability of BWMS are ambient water salinity (the ability to treat fresh, brackish, and marine water) and temperature (the ability to work effectively and safely in a variety of temperatures from warm equatorial to cold polar water), ship ballasting rate (the ability to treat water moving at a variety of flow rates from < 200 m^3/hr to > 4,000 m^3/hr), and ballast volumes (the ability to treat total volumes of ballast water from < 1,000 m^3 to > 50,000 m^3).

Another important vessel consideration is impacts of treatments on ballast tank and piping coatings and substrate corrosion rates. Nearly all systems that alter the chemical composition or reactivity of ballast water (e.g., heat, oxidants, and deoxygenation) can potentially affect corrosion of ship structures, piping, fixtures and protective coatings. To a great extent, the potential effects of these BWMS have not been consistently evaluated across the various modes of corrosion, including uniform or localized corrosion, or for potential interactions with corrosion control systems including protective coatings and cathodic protection systems. Some BWMS have provided data that indicate negligible impacts on corrosion rates or even improvements. For example, deoxygenation, if operated properly, can dramatically reduce uniform corrosion rates, but alternatively, deoxygenation may result in increased corrosion rates due to either the cycling of hypoxic and aerated conditions or the formation of corrosion-causing sulfate reducing bacteria if anoxic conditions are reached. Similarly, other BWMS utilizing strong oxidants have been evaluated as having apparently negligible effects on coatings and steel corrosion rates. However, it is also well documented in the water treatment and marine vessel industry that continuous exposure to high doses of some oxidants, such as halogenated oxidants, can cause severe corrosion rates (depending on the specific oxidant, its concentration and contact period). On the other hand, while heightened corrosion rates may be experienced shortly after treatment, corrosion rates on the whole may not be significantly affected if the oxidant concentration declines rapidly.

Corrosion is already a significant concern for vessels operating in saltwater environments. As such, coating failures and steel wastage are currently incorporated into periodic surveys and vessel service periods. In the end, an increase in corrosion rates will impact the maintenance and repair costs borne by the vessel owner; these potential increases in cost will need to be factored by the owner in selecting a BWMS. In addition, corrosion control and mitigation strategies such as coatings and cathodic protection should also be carefully considered since either or both of these may be employed to offset any increased corrosion concerns. Although comprehensive assessments have not been conducted for all BWMS, no major damage or casualties related to corrosion have been identified to date for BWMS installed on ships.

In addition to specific environmental and vessel applications, vessel type and vessel operations can dictate BWMS applicability. Although a multitude of vessel designs and operation scenarios exist, a few important examples of specific constraints can greatly limit treatment options. Perhaps the most dramatic limitations are found with the Great Lakes bulk carrier fleet that operates vessels solely within the Great Lakes with large volumes of fresh, and often cold, ballast water ("Lakers"). The vessels in this fleet have ballast volumes up to 50,000 m^3, high pumping rates (up to 5,000 m^3/hour), uncoated ballast tanks (older vessels), and some vessels have separate sea chests and pumps for each ballast tank. A further confounding issue is that voyages taken by Lakers average four to five days, with many less than two days. Given these characteristics, a number of limitations are imposed: electrochlorination and ozonation may only work in freshwater with the addition of brine (in particular Cl and Br, respectively); oxidizing chemicals may increase the corrosion rate of uncoated tanks; deoxygenation and chemical treatments that require holding times to effectively treat water (or for the breakdown of active substances) may not be completely effective on short voyages; and the space and power needed for the required numbers of filtration + UV treatments may simply not be available.

Another example of vessel-specific constraints is the sheer size of some vessels and the cargo they carry. Very Large Crude Carriers (VLCC) and Ultra Large Crude Carriers (ULCC) can carry up to 100,000 m^3 of ballast and can fill or discharge ballast water at over 5,000 m^3/hour. While various BWMS may be modular (perhaps providing the ability to add several units in a manifold design or in sequence), systems that include a mechanical separation stage (e.g., filtration, hydrocyclone) or exposure to UV or sonication may have difficulty addressing these large volumes and flow rates. Furthermore, given the hazardous nature of the cargo carried on these ships (and other similar vessels, such as Liquefied Natural Gas carriers), restrictions on the placement of a specific BWMS may apply, and system components will likely have to satisfy classification society requirements for explosion-proof and intrinsically safe construction, which might be more difficult for some BWMS types than others.

A final example is the treatment of ballast water on the tens of thousands of unmanned barges in the U.S. that would fall under the ballast water discharge regulations. Inland waterways and coastal barges are not self-propelled, but rather are moved by towing or pushing with tugboats. Because these vessels have been designed to transport bulk cargo, or as working platforms, they commonly use ballast tanks or fill cargo spaces with water for trim and stability, or to prevent excessive motions in heavy seas. However, the application of BWMS on these vessels presents significant logistical challenges because they typically do not have their own source of power or ballast pumps and are unmanned.

Conclusion: While several BWMS appear to safely and effectively meet IMO D-2/Phase 1 discharge standards, there are several factors to consider beyond mechanical and biological efficacy. A variety of environmental (e.g., temperature and salinity), operational (e.g., ballasting flow rates and holding times), and vessel design (e.g., ballast volume and unmanned barges) parameters will impact the performance or applicability of individual BWMS.

5. System Development

5.1. Introduction

This section addresses issues related to potential future improvements in BWMS in response to charge question 3, which focuses on three main issues: improving the performance of existing BWMS technologies, identifying impediments to improved technological performance, and considering whether technologies can achieve zero or near-zero discharge of organisms.

Before considering possible improvements, it is important first to consider what has been achieved to date by existing BWMS technologies. The 2004 IMO D-2 standard has provided a stable target for research, development, testing, and evaluation of practices and technologies to treat ballast water. Using the proposed D-2 standard as a design goal, some developers of BWMS have:

- Integrated BWMS within marine vessel arrangements, weight and stability constraints, electrical distribution and piping systems, and automated control systems.
- Integrated the operation of BWMS within the larger context of merchant vessel operations such as ballasting rates and volumes, logistics requirements such as reliable chemical-supply chains and service/support centers, safe operations such as hazardous-rated equipment and chemical-handling procedures, and operational training.

- Tuned BWMS to achieve acceptable levels of disinfection by-products, residual toxicity, within the limits of practicality and in compliance with the proposed D-2 standard.
- Packaged the technology for a competitive commercial market, which requires consideration of life-cycle costs, reliability of equipment, maintenance issues, and acceptance of the BWMS technologies by mariners.

These are important achievements. They also foreshadow the issues that must be considered when evaluating technological options to improve BWMS technologies.

In general, technological changes to improve BWMS performance will proceed along one of two paths: (1) incremental changes to existing designs with the goal of optimizing performance, or (2) designing entirely new treatment methods. Incremental changes offer the faster path to improved performance, but are likely to achieve only relatively modest improvements. Wholly new approaches for BWMS, possibly drawn from the water treatment industry, also would improve performance—perhaps significantly—but would take more time to develop and test in order to determine performance, practicality, and cost- effectiveness. The Panel considers both pathways below.

5.2. Improving the Performance of Existing Systems

Charge question 3a. For those systems identified in questions 1.a and 2, are there reasonable changes or additions to treatment processes which can be made to the systems to improve performance?

The Panel defined “reasonable changes” as incremental adjustments or improvements that do not fundamentally alter the treatment process. For example, design changes to increase UV radiation intensity would be an incremental adjustment, whereas addition of a UV stage would not. In practice, “reasonable changes” mean the same thing as “incremental improvements.” Both are based on the concept of “turning up the dial” on existing technologies rather than creating wholly new systems or adding processes to the treatment system. Although incremental changes can generally be implemented more quickly, this approach is not necessarily simple or foolproof. First, it may be impossible or impractical to further improve the baseline technology. Second, any changes could fundamentally alter other aspects of the technology’s performance or use; e.g., it could change its life cycle costs, affect integration of the BWMS with vessel operation, or increase residual toxicity of ballast water discharges.

As described in Section 4.4, five BWMS have demonstrated compliance with IMO D-2 standards, under G8 testing conditions. Based on information from available test results, incremental improvements could be made to these treatment processes, perhaps yielding performance greater than D-2. However, these changes may add costs and engineering complexity.

Examples of incremental improvements to the BWMS judged to comply with the D-2 standard, as shown in Table 4-1, are summarized below.

- **Deoxygenation + cavitation**. Current technology for these systems establishes severe hypoxia, which kills larger organisms very effectively, although hypoxia has little or no effect on some bacteria, pathogenic protozoa, or viruses. It is not possible to improve on hypoxic conditions *per se*, however, it may be possible to reduce the time needed to reach severe hypoxia and increase holding time under severe hypoxia. In addition, effectiveness might be improved by increasing the degree of cavitation and physical/mechanical disruption of organisms.
- **Mechanical separation + oxidizing agent**. These systems could be optimized in several ways: improved mechanical separation (i.e., filtering) to remove higher percentages of particles and particles of smaller size; increasing the concentration and contact time for oxidizing agents; and adjusting other water chemistry parameters, such as pH, to increase the efficacy of the oxidizing agent. (Note: this category corresponds to those BWMS using filtration and chlorine dioxide, filtration plus UV + TiO^2, and filtration plus electrochlorination).
- **Mechanical separation + UV.** These systems could be optimized by improved mechanical separation (i.e., filtering) to remove higher percentages of particles and particles of smaller size and by increasing UV contact time and dosage.

5.2.1. Combination Technologies

Most ballast water treatment systems, even those with a single primary component, are actually combination technologies. For example, one company's BWMS relies primarily on deoxygenation, but also has a venturi device that mechanically damages some of the organisms (as would cavitation) and uses carbon dioxide, which forms carbonic acid, lowering the pH of the water. Another commercial system is advertised as a combination technology that includes filtration, ultraviolet radiation, and treatment by free radicals.

It is difficult to fully understand the interactions of treatment processes used in combination BWMS. This makes it hard to predict the overall treatment effect of incremental improvements within individual processes. For example, one company's system combines filtration, cavitation, ozone, and sodium hypochlorite. With four "primary" technologies at work, which should be the focus for "turning up the dial" to improve performance? Further complicating matters is the great variability in the physical and chemical properties of ballast water itself, which in turn creates complex interactions with individual treatments as well as their combination.

To date, combination BWMS have been developed through research and testing. However, once a technology has shown promise to meet the D-2 standard, its development is stopped to allow the device to undergo certification testing. It is reasonable to assume that incremental improvements to combination technologies could yield efficiencies in operation (less power, less cost, more reliability) and moderate improvements in treatment effectiveness. Due to the complex interactions among combined treatment processes, however, such possibilities can only be speculative until a more rational understanding of the modes of inactivation/kill is developed and verified by experiments with prototypes. Thus, the Panel's comments on likely improvements to BWMS technologies are restricted to the primary treatment processes identified in Section 4 and described below: UV radiation, mechanical separation + cavitation, deoxygenation, and oxidant based systems. Other BWMS processes, such as ultrasound and electro-mechanical separation, are not as widely utilized and therefore were not reviewed.

5.2.2. UV Radiation

There are several ways in which treatment by UV radiation may be improved. It may be possible to deploy “over-sized” UV radiation treatment systems to improve performance. For example, a ballast system that runs at 800 m^3/hr could be paired with a treatment system rated for 1,000 m^3/hr, thereby presumably increasing UV exposure by 20 percent. Testing and analysis would be required to determine if efficacy was actually increased and, if so, by how much, and to ensure there were no adverse impacts to the residence time distribution or UV intensity field distribution of the UV chamber.

Similarly, UV system performance might be improved through increased intensity of UV lamps. The length of time the ballast water is exposed to UV radiation could also be increased by increasing the size of the chamber relative to the ballasting rate. Such improvements, however, would increase the size and cost of BWMS equipment.

UV chambers could also be staged in series, though this would substantially increase cost, required space, and maintenance. However, employing multiple UV chambers in series could provide the following improvements to performance:

- Decreased chance that organisms could “slip” past untreated, assuming that each chamber could independently provide treatment adequate to meet a given standard.
- Increased time during which organisms are exposed to UV.

The performance of UV systems also could be improved by using more effective mechanical separation methods (“filtering”) upstream of the UV chambers in order to enhance the transmissivity (clarity) of the ballast water prior to UV treatment. Flocculants such as alum could further clarify the ballast water, providing that these agents do not impart UV absorbance (as may be the case with iron-based flocculants). Such improvements have drawbacks. For instance, advanced mechanical separation would significantly increase costs and space requirements, and would likely significantly increase backpressure within the system, resulting in higher electrical power demands and a need for higher-head ballast pumps.

5.2.3. Mechanical Separation and Cavitation

Many BWMS use mechanical separation as the primary precursor for other treatments such as UV radiation or oxidants. The purpose of mechanical separation varies according to the treatment’s disinfection processes: e.g., screening to remove larger organisms resistant to disinfection, reduction of organic matter to reduce oxidant demand, and reduction of turbidity to increase transmittance of UV radiation.

Mechanical separation also has a secondary effect—physically damaging some of the organisms as they pass through the device—which may inactivate or kill organisms or weaken their cellular structure such that effective disinfection is more easily achieved. In this regard, mechanical separation is similar to cavitation devices designed to impart physical damage to organisms. Some BWMS include cavitation devices to damage cellular structure without having to handle separated filtrate.

Use of seawater filtration on vessels traditionally has been limited to protecting mechanical devices in the piping system. For example, seawater might be “screened” to a one-eighth inch opening (3.175 mm) to protect the narrow passages of a heat exchanger.

Recently, however, several common and proprietary devices have been developed for filtering and imparting cavitation effects on ballast water as part of the treatment process: variations on back flushing of traditional screen filters; vibrating disc filters; multi hydro-cyclone; and various cavitation devices. In general, the filter units target removal of particles above 40 or 50 μm and have significant waste streams that are returned to the ambient water. Typically, filtering takes place on ballast water uptake only.

Mechanical separation devices are advertised in terms of percentage removal. For example, two companies claim filtration rates of approximately 90 percent removal of zooplankton. These removal levels, although essential to support the disinfection process, by themselves are far from adequate to meet the D-2 standard for the size class > 50 μm.

In summary, it is not reasonable to expect incremental improvements in mechanical separation devices to achieve significant advances in performance over the D-2 standard. Such improvements would require use of media filters, membrane filters, or other devices that have not yet been practically applied to ballast water treatment. Similarly, cavitation devices alone cannot meet the D-2 standard. It is not clear if improvements to cavitation devices will significantly increase the effectiveness of BWMS that employ combined processes.

5.2.4. Deoxygenation

Two Type-Approved BWMS that have met the D-2 standard use deoxygenation as part of multiple treatment processes. The first system lowers oxygen by pumping low-oxygen gas from a purpose-built burner into the BWMS through a venturi device. The efficacy of this system relies on several interrelated components: rapid application of the gas stream; creating carbonic acid from the carbon dioxide in the gas stream, which lowers pH and makes the low-oxygen environment more lethal; and the mechanical effect of the venturi on passing organisms.

This system lowers the oxygen level to about 2 percent by volume (about 0.7 mg/L) by using a variation on the traditional tank-ship combustion-based inert-gas generator. Traditional units produce a 5 percent oxygen level (about 1.8 mg/L). For reference, 2 mg/L oxygen is considered the upper boundary for environmental hypoxia and the point of mortality for sensitive species. Very few metazoans can survive <1 mg/L oxygen for longer than 24 hours (Vaquer-Sonyer and Duarte 2008). Further improvements to combustion-based units may not be practical given constraints of the combustion process.

The second system lowers oxygen levels through use of a nitrogen generator. The generator uses a membrane to filter ambient air, resulting in high quality nitrogen gas. The system also uses mechanical separation, cavitation and electrodialytic disinfection. Nitrogen generators are widely deployed in industry and in some marine applications. On ships, they are generally regarded as expensive, high- demand consumers of electrical power. It is possible to create very high quality nitrogen gas, approaching 99.9 percent pure, but doing so requires significant space, capital costs, and high electrical power demands.

As these examples show, BWMS that use deoxygenation often also use additional treatment processes. Thus, it is difficult to predict the effect of incremental improvements on overall efficacy. In fact, some changes might decrease efficacy, or worse, result in unanticipated adverse conditions; e.g., higher populations of sulfate-reducing bacteria and subsequent increase in steel corrosion rates. Consequently, it is not possible to assess whether incremental improvements will yield higher performance. For example, with respect to treatment lethality to metazoans, there is little to no difference between oxygen levels of 2 mg/L and 1 mg/L for the same contact period. Extending holding time would be more efficient than additional efforts to reduce oxygen below 1 mg/L. However, there is some evidence (Mario Tamburri, pers. comm.) that faster transitions to severe hypoxia are more lethal.

5.2.5. Oxidant-Based Systems

Oxidant-based systems introduce an oxidizing agent (such as chlorine) into the ballast water stream. These systems are generally designed to target a given level of total residual oxidant (TRO) in the treated ballast water. Oxidant-based systems pose issues for mechanical integrity and for worker safety, since these systems require adding chemicals in bulk, on-site manufacture of sodium hypochlorite or similar chemicals, or on-site production of ozone gas.

The consumption (or oxidant demand) of the introduced oxidant varies with the organic-matter content of the ballast water uptake. After the initial instantaneous consumption of oxidant, any remaining oxidant is pumped with the ballast water into the vessel's ballast tanks. There it is held for a prescribed length of time at the TRO concentration. The TRO level will decay over time as a function of many factors, including its initial concentration, salinity, temperature, motions of the vessel, and configuration of the ballast tank and venting system. Depending upon the predicted or measured oxidant levels in the ballast water, a neutralizing agent may be applied before or during its discharge to the environment.

The effectiveness of oxidant-based systems is a function of the concentration of the residual oxidants and the holding time. Incremental changes that could improve effectiveness include: increasing initial oxidant concentrations; maintaining a higher oxidant concentration during the holding period; and increasing the holding period or contact time. The potential for improved performance for each of these three options are considered below (the use of oxidants in combination BWMS treatments is considered separately).

Increasing Initial Oxidant Concentrations

Determining the initial oxidant concentration needed to reach the required efficacy is part of the "art" of a BWMS. For example, TRO values for BWMS that the Panel reviewed varied in the type of oxidant used (e.g., ozone, chlorite ion, free active chlorine) and TRO amounts varied by an order of magnitude.

Several oxidant-based systems also use some form of mechanical separation to remove larger organisms, organisms entrapped within protective solids, and some particulate organic matter and thereby reduce oxidant demand. Regardless of the effectiveness of the mechanical separation, however, it is the residual oxidants and other reactive disinfection byproducts that achieve the final treatment performance. Thus, even though mechanical separation may reduce the amount of chemical required, it is unlikely to improve the efficacy of the oxidant. Tertiary effects also occur, such as damage to organisms' membranes during mechanical separation and subsequent membrane interaction with oxidant-based systems. However, these tertiary effects are difficult to assess. Thus, they do not represent an obvious method for incremental improvement.

However, it is possible to "turn up the dial" on existing BWMS by increasing the amount of oxidant used, which should improve effectiveness. Doing so simply requires that a higher capacity ballast water treatment system be installed. For example, concentrations could be increased 50 percent by installing a system rated for 1200 m^3/hr on a vessel that pumps ballast water at 800 m^3/hr. Such an installation will demand larger space and weight allowances, more power, and higher capital and operating costs. In general, it should be possible to integrate higher capacity systems for new vessel designs, but it is more of a challenge to retrofit on existing vessels.

Higher oxidant levels in ballast water can have a significant negative effect impact on piping-system components and tank-coating systems. Valve packing, flange gaskets, and pump seals are made of a variety of materials, some of which are not compatible with oxidants at low concentrations, and less so at increasingly higher concentrations. Impacts on tank coatings are not yet well understood. TRO levels up to 10 mg/L may be compatible with typical intact ballast-tank marine coatings. However, coatings are frequently not intact, because they wear over time. In the case of freshwater shipping, a ballast water tank may not be coated at all. Corrosion of exposed carbon-steel structures can lead to structural failures and require expensive and complex repairs. Use of increased oxidant levels, therefore, would likely increase rates for coating failures and corrosion of exposed carbon-steel structure.

Use of higher oxidant levels also increases concerns for safe handling on board vessels, due to the need for handling and storage of additional bulk chemicals, lengthening the time required to make confined tank spaces safe for entry for inspection and repair work, and generation of hydrogen gas. These concerns can be handled through operating procedures but at the expense of increased time and effort. As higher levels of oxidants are introduced into ballast water, complex chemical reactions take place, resulting in potentially harmful disinfection byproducts through interactions between the oxidant level and characteristics of the uptake water (such as its organic load, alkalinity, salinity, and chemical contaminants). Further tests and analysis would be required to determine whether these byproducts should and can be neutralized so that the ballast water discharge will meet acceptable toxicity limits.

Maintaining Increased Oxidant Concentrations

Most oxidant-based systems rely on achieving residual oxidant levels that are adequate to meet D-2 standards and then maintaining that concentration for the duration of the holding period. The hold time of ballast water can vary significantly, however, and shipping schedules, weather, equipment failure, and cargo-handling changes frequently result in hold times that are longer or shorter than initially expected. As hold times increase, TRO concentrations decay, which also reduces detoxification costs. Most evaluation testing occurs during a prescribed holding period, typically for two to five days. In reality, ballast water hold times routinely vary from one day to several weeks. In fact, some ballast tanks can remain full, or partially full, for many months or even years.

There has been little development or testing of systems that monitor and maintain a specific oxidant level in ballast water tanks. Indeed, automated monitoring of oxidant levels in ballast water tanks is not currently practiced. Continuous or periodic monitoring would require either a network of sensors installed in the tanks or a means of drawing a liquid sample on a periodic basis to a remote monitoring device. Such sensors are common practice in the treatment of drinking water and wastewater. Either approach requires significant cabling, possibly tubing and pumps, monitoring equipment, and data- recording devices.

Current practice to maintain oxidant levels, if done at all, is to "top up" a ballast tank; i.e., to partially discharge its contents, then refill the tank with freshly treated water. The objective is to achieve the desired oxidant level by mixing the "new" water having a high concentration of oxidant with the water remaining in the tank. Under such conditions, it is imprecise to determine whether desired inactivation levels are achieved. Such efforts are similar in mechanical function to ballast water exchange, would likely be performed while the vessel is at sea, and carry with them the same significant safety concerns regarding vessel stability. A safer and more reliable approach for topping up oxidant levels would require developing new systems. Such systems might include chemical dosing lines to deliver an external supply to each ballast tank, combined with circulation devices internal to each ballast tank.

Increasing the Hold Period

Increasing the hold time of the ballast water while maintaining a certain oxidant level would likely increase treatment efficacy. However, it is ship operations that will dictate the duration of this hold time for most ballast water tanks. In particular, the largest mid-body ballast water tanks almost always have to be discharged while tank ships or bulk carriers are being loaded. As such, the treatment process must account for the expected hold period, but likely will not have the ability to alter it.

Summary of Oxidant-Based Treatments

Existing oxidant-based systems have been developed to meet the D-2 standard, and several have received international approvals. Their efficacy could be improved by increasing initial residual oxidant levels in ballast water during uptake. However, testing would be needed to determine the degree of improvement and to determine toxicity effects from disinfection byproducts resulting from higher oxidant doses.

Increasing residual oxidant levels also creates demands for space, weight, power, and capital and operating expenses; in addition, these systems will increase piping-system compatibility issues, ballast-tank corrosion rates, and safe-handling concerns. Alternatively, it may be possible to increase effectiveness by maintaining residual oxidant levels during holding time in the ballast water tanks. Current systems, however, have only rudimentary methods for performing such operations. New methods will need to be developed and tested to determine their practicality and effect.

5.3. Principal Technological Constraints

Charge question 3b. What are the principal technological constraints or other impediments to the development of ballast water treatment technologies for use onboard vessels to reliably meet any or all of the discharge standards?

Existing BWMS have been developed to the IMO D-2 standard within the context of typical marine vessel constraints, including restrictions on size, weight and energy demands. While practical for new construction vessels, existing vessels may not be able to integrate such BWMS on a retrofit basis. Meeting higher standards generally implies that the treatment processing plants will need to be large, heavy, energy-intensive and expensive. At some point, constraints associated with the installation and operation of such equipment may require a fundamental shift in how ballast water is managed.

Regardless of the applicable treatment standard, existing and potential BWMS share common impediments to development.

(a) Technical constraints include:

- The shipboard marine environment is corrosive and subject to vibrations and ship motions. Thus, one should not assume that shore-side systems can be transferred easily to shipboard use; in fact, a strong shipboard service history will be an important guide to selecting system components. Even so, the characteristics of water in some non-ballast water shipboard applications may differ from ballast water (e.g., sediment concentrations may be greater in ballast water). This makes it difficult to predict performance based on service history alone.
- Vessels are initially designed to have ballasting capabilities and procedures that meet their intended service and voyage profile. BWMS intended for retrofit will need to fit within those original parameters.

(b) Other impediments include:

- *Lack of clear design goals.* There is disagreement on discharge standards; they vary from state to state within the U.S. and internationally. Thus, BWMS manufacturers have multiple discharge standards as design targets. Further, there are no established compliance monitoring, enforcement procedures. Such procedures would help focus future BWMS development; e.g., they would encourage the creation and use of additional performance metrics, such as system reliability, as contrasted with the current focus on discharge quality as evaluated during certification testing.
- *Limited experience and limited empirical data on life-cycle costs.* The full cost of any BWMS includes not only its initial purchase and installation costs, but also its long-term operational costs. System reliability, durability, cost of spares, and ease of maintenance are factors that contribute to determining the BWMS life-cycle costs.

(c) Constraints to improved performance include:

- Shipboard BWMS are developing rapidly. The focus to date has been engineering the treatment device for discharge performance. This focus has come at the expense of ensuring integration of the BWMS with vessel mechanical systems and marine operational activities; BWMS durability, maintenance and repair; training; and procedures for monitoring technology performance.
- Ships crews are small in number and busy; therefore, any new system must be easy to operate and maintain. Ideally, new systems would enable remote control from the ballast control console and automatic operation in or near port, which is typically a busy time for crew.
- Most importantly, BWMS should pose no unreasonable health risk for the crew nor create higher risk for vessel safety, and require no exceptions to the safety procedures established by the vessel owner. The BWMS installation and operation procedures must also meet the requirements of control authorities, i.e., Classification Society, Flag State, and Port State.
- Facilities properly equipped to test BWMS technologies are few, which imposes a bottleneck to swift verification and testing and thus hinders development. Increased sharing of data and specific protocols among such facilities is essential.

5.3.1. Operational Challenges on Working Merchant Vessels

It is unlikely that current BWMS will be able to meet the most stringent proposed standards (e.g., 100x D-2 or 1000x D-2). This is perhaps best understood in the context of required reductions in organisms. Meeting the D-2 standards for zooplankton-sized organisms requires that the BWMS reduce the number of zooplankton in challenge water (as defined by the EPA ETV program) by four orders of magnitude. For a very large crude carrier (VLCC) tanker carrying roughly 90,000 m^3 of ballast water, the D-2 standard would require reducing the number of zooplankton-sized organisms from 9 billion to 900,000 (Table 5-1). The USCG's proposed Phase 2 standard for zooplankton (in the column labeled "D-2/1000" in Table 5-1) would require that BWMS reduce viable zooplankton by seven orders of magnitude relative to values in ETV challenge water. (This is a 99.99999 percent reduction, referred to in reliability engineering as "seven-nines"). For the VLCC example, the proposed Phase 2 standard would limit the discharge of viable zooplankton to a maximum of 900 individuals (Table 5-1). To put this value in perspective, it is fewer than half the number of zooplankton (2000 individuals) contained in a 20-liter bucket of ETV challenge water (Table 5-1). Additional examples of allowable zooplankton discharges associated with different discharge standards and for different types of vessels are summarized in Table 5-1, below.

In contrast, incremental adjustments to existing technologies are expected to result in only slightly greater reductions of viable organisms in BWMS discharge. In part, the inability to achieve huge reductions stems from the design characteristics of present-day BWMS technology, which is placed "on top of" existing ballast-piping systems. The treatment devices (e.g., filters, UV lamps, cavitation devices) are added to the standard ballast-piping, which was originally designed solely for the efficient uptake and discharge of ballast water. Further, ballast water is still taken up, held, and discharged in essentially the same manner as in the past.

It is also instructive to consider the challenges of meeting the proposed, more stringent standards within the context of a working merchant vessel. Table 5-1 shows that VLCC tankers discharge ballast water at a rate of 5,000 m^3/hr. At this rate, one second of discharge yields 1.39 cubic meters of water. Assuming ETV challenge water conditions, this one second of discharge would contain 139,000 zooplankton – a number that would exceed the allowable discharge of organisms for the entire VLCC ballast water capacity by the following amounts for the proposed more stringent standards: 1.5 times greater than the 10x D-2, 15 times greater than the 100x D-2, or 154 times greater than the 1000x D-2. Meeting these more stringent standards will require the following technical challenges be overcome:

Controls to Avoid Discharge of Untreated Water

At a minimum, ballast water piping systems must be carefully designed to avoid discharge of any untreated ballast water, however minimal in volume. Doing so likely would require separate uptake and discharge ballast water piping. Current standard practice is to use a common piping system for both uptake and discharge. In addition, guarding against discharges during brief interruptions in treatment during start-up or shut-down may require that BWMS be designed to re-circulate treated ballast water to confirm its treatment status before discharge.

Table 5-1. Zooplankton Counts for Water and Increasing Log Reductions from D-2 Standard. The USCG's proposed Phase 2 standard is represented by in the column labeled "D-2/1000"

Volume Basis	Volume *(m3)*	Rate *(m3/hr)*	*Viable Organisms >50 um (Seawater per US ETV)*				
			Seawater	IMO D-2	D-2 x 10	D-2 x 100	D-2x1000
Test Standards	1.00E+00	NA	1.00E+05	1.00E+01	1.00E+00	1.00E-01	1.00E-02
VLCC Tanker	9.00E+04	5.00E+03	9.00E+09	9.00E+05	9.00E+04	9.00E+03	9.00E+02
Great Lakes Bulk Carrier	4.40E+04	1.00E+04	4.40E+09	4.40E+05	4.40E+04	4.40E+03	4.40E+02
Handymax Bulk Carrier	1.80E+04	1.30E+03	1.80E+09	1.80E+05	1.80E+04	1.80E+03	1.80E+02
Panamax Container	1.70E+04	5.00E+02	1.70E+09	1.70E+05	1.70E+04	1.70E+03	1.70E+02
Feedermax Container	3.50E+03	4.00E+02	3.50E+08	3.50E+04	3.50E+03	3.50E+02	3.50E+01
Passenger Ship	3.00E+03	2.50E+02	3.00E+08	3.00E+04	3.00E+03	3.00E+02	3.00E+01
ETV Testing Tank	2.00E+02	2.00E+02	2.00E+07	2.00E+03	2.00E+02	2.00E+01	2.00E+00
VLCC Pipe (2.2 meters)	1.39E+00	5.00E+03	1.39E+05	1.39E+01	1.39E+00	1.39E-01	1.39E-02
Bucket (20 liters)	2.00E-02	NA	2.00E+03	2.00E-01	2.00E-02	2.00E-03	2.00E-04
Glass (0.4 liters)	4.00E-04	NA	4.00E+01	4.00E-03	4.00E-04	4.00E-05	4.00E-06

Table 5-1 expresses zooplankton treatment standards as maximum allowable numbers of viable organisms for various volumes. The top row ("Test Standards") provides organism counts in 1 m3 of ETV challenge water (column labeled "Seawater"), and maximum allowable counts in 1 m3 of water meeting the IMO D-2 standard and successive log reductions beyond D-2. Several vessel types are listed showing their typical ballast-water volumes and discharge flow rates. For each volume, the table shows the number of organisms it contains (column labeled "Seawater") and the maximum number of organisms allowed by each of the discharge standards.

Table 5-1 also indicates the number of zooplankton in volumes of ETV challenge water equivalent to a beer glass, a bucket, and that displaced by one second of untreated discharge from a VLCC. The colored highlights indicate when the glass, bucket, or discharge contains more viable organisms than those in the total volume of water discharged from a vessel in compliance with the various standards.

Controls to Avoid Cross-Contamination

- *Isolating the ballast-piping system.* Many ships have a cross-over to fire mains, black and grey water drains, bilge water lines, and cooling water circuits.
- *Maintaining a high level of tank structure integrity.* Especially in aging vessels, tank structures can permit transfer of fluids from adjacent tanks, piping systems running through tanks, fluids pooling on tank tops, and directly from ambient water through seams or pipe fittings in the vessel's side shell.
- *Protecting tank vents.* Ballast tanks vents are typically fitted with only a rough screen or a ball check device to minimize entry of seawater. Protecting these vents from ingress of untreated seawater will become more critical if standards become more stringent.

In-tank Monitoring, Treatment and Mixing

Careful monitoring of in-tank conditions will be very important under the following conditions: when ballast water hold times are very long, thus enabling surviving organisms to reproduce, or when they are very short, thereby reducing time for treatment to take effect; when in-tank sediment loads provide a protective layer for organisms, shielding them from the disinfection process; and when highly heterogeneous ("patchy") uptake ballast water overwhelms the treatment process.

Overcoming these challenges requires developing means to:

- *Monitor tank conditions*, although doing so is difficult because ballast water tanks are typically complex and are known to have hydrodynamic "dead zones" that are not flushed out during a typical ballast cycle.
- *Treat (or re-treat) a full ballast water tank*, such as would be needed when the ballast water uptake is ineffective, when it has been contaminated from external sources, or when the expected hold time has been exceeded.
- *Mix a full ballast water tank.* An ideal mixing system would suspend sediment loads, permit even treatment of the tank's entire contents, and permit representative monitoring of the tank.

Improving the Efficacy of Mechanical Separation and Disinfection Technology

The performance of current "filter and disinfect" treatments is especially limited in circumstances when the ballast water uptake is patchy or has a high sediment load.

5.3.2. Idealized Designs for BWMS

The water treatment industry is an obvious place to turn for developing new BWMS. This industry has developed methods to disinfect large volumes of water to very high standards. New approaches adapted from that arena may be very efficacious and able to achieve the proposed, more stringent standards, but it would take time to develop, test, and determine their practicality and cost impacts. Nonetheless, in thinking about an idealized design for ships, it is a useful thought exercise first to consider elements from a shore-based treatment system.

To that end, the Panel developed a hypothetical design for an onshore ballast water treatment plant with a design capacity of 20,000 m^3 of ballast water per day. This is equivalent to ~800 m^3 per hour, roughly similar to a "low ballast dependent" vessel such as a containership. ("High ballast dependent" vessels, such as Great Lakes bulkers and large tank ships, would require a treatment plant 5 to 12 times larger.)

The design requirements for this idealized, hypothetical treatment plant were estimated as:

- Equalization tanks of volume 20,000 m^3.
- Plain sedimentation area of ~1,000 m^2.
- Granular media filtration of ~120 m^2.
- Three UV units each at ~800 m^3 per hour.
- Sludge and backwash handling.
- Possibly to include a membrane-filtration unit.

Based on the long history of water treatment plants, the Panel thinks it likely that such an idealized system could meet IMO D-2, and indeed, 1000x D-2 standards for all size fractions, including the IMO- specified bacteria. Nonetheless, pilot scale testing would be needed to confirm optimum design parameters. Such pilot testing programs are common practice in water treatment plant design.

Using the Idealized Design as a Basis for Conceptualizing New Shipboard BWMS

The previous section describes opportunities for incremental improvements in BWMS. Here the Panel illustrates the second pathway for improved BWMS – that is, the design of wholly new systems – for the purpose of meeting the proposed, more stringent standards. The example concept discussed below draws upon the operational particulars of the idealized system just described within the context of the technical and operational constraints for shipboard BWMS.

A wholly new treatment design would significantly increase the operational burden on ship operators, but it is technically feasible to integrate wholly new treatment systems into *new* vessel designs. Integrating such an idealized system into *existing* vessels would be technically challenging on most, and not possible on many, existing vessels. Finally, such a conceptual system and processes would need better definition and specification in order to develop cost-benefit analyses; neither capital nor operating costs have been estimated.

In order to better convey the distinction between incremental improvements to BWMS and new designs for BWMS, an illustrative, hypothetical concept sketch of a new design for shipboard BWMS is presented in Figure 5-1. This concept sketch is based on a Panamax container ship having a ballast volume of 17,000 m^3 and a discharge rate of 500 m^3/hr. This sketch is illustrative only. It is presented solely to assist in the evaluation of how more stringent treatment standards might impact vessel arrangements, operations and costs. For example, this system would likely require at least three to four times the number of components, space, expense, and effort compared to existing BWMS.

By way of overview, this conceptual system would be capable of achieving higher filtration levels, provide greater control of oxidant levels in tanks, and enable a final disinfection using UV radiation. The treatment process would be integrated through use of large media filters integral to the vessel hull for ballast water uptake and discharge and through recirculation of the ballast water in the ballast water tanks in order to dose, monitor, and maintain oxidant levels in the ballast water tanks. For ballast water discharge, a residence tank would be considered adequate to ensure neutralization of the oxidant. A final UV disinfection step would be handled using a dedicated ballast water discharge connection. Details for each of these steps for this idealized system are described below.

Ballast Water Uptake

Two traditional, but oversized, sea chests (intake structures for ballast water in ships' hulls) would serve to take up ballast water. Piping would generally be 300 mm nominal. Each sea chest would include standard skin-valve isolation and piping materials. The sea chests would be located port and starboard, one high and one low, with a cross-over suction main connecting each. This would provide flexibility for avoiding sediment when the ship is close to the bottom, and algal blooms when the ship is light and the high sea chest is close to the surface. The Panel recognizes that sea chests can provide refuge for nonindigenous species, but methods for keeping them and adjacent hull areas free of fouling organisms were not considered, as such considerations were beyond the charge.

The cross-over suction main would discharge by gravity into two large media chambers plumbed in parallel and each sized for full flow. This arrangement would allow one to be by-passed during back- flush cycles. Each would be built into a one-meter-height double bottom in the ship's hull and 8 m^2 for a volume of 64 m^3 each. Industrial wastewater industry media with tolerance for velocities approaching 60 m/hr, and a useful life of six years between dry dock periods would be considered. Six-year servicing of media would be through manhole covers.

Ballast water leaving the media filter would be disinfected prior to entering the ballast water tanks, either by a UV or an oxidant chemical. This transfer would be possible by using ballast water pumps, or through gravity when there is adequate head pressure from the sea. The piping would be direct, through a pipe tunnel for ease of monitoring condition and servicing, and have no cross-connects.

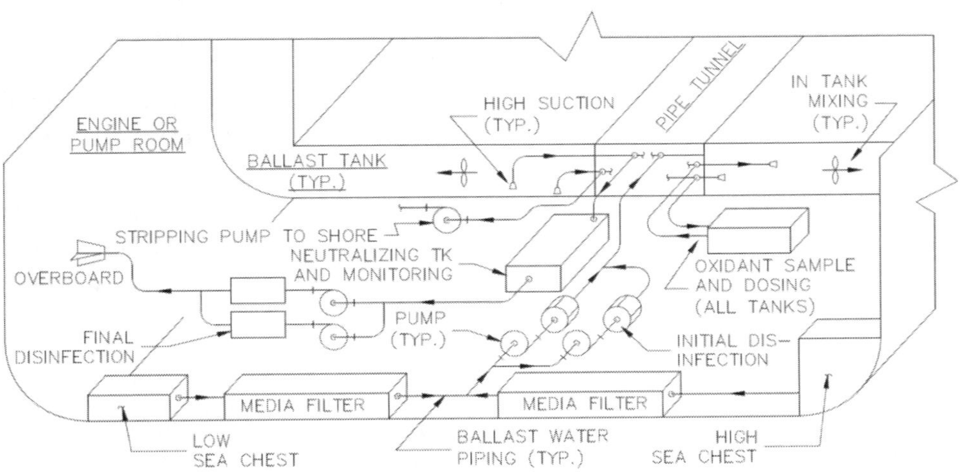

Figure 5-1. Concept Sketch of a New Approach to Shipboard Ballast Water Treatment (TYP means "typical").

In Tank

Once a ballast water tank is full or partially full, it would be periodically mixed through the use of low- pressure high-volume air bubbles, or in-tank eductors. This mixing would allow the application of an oxidant to a prescribed level, and the monitoring and the maintenance of that oxidant level. Mixing frequency would be based on detected oxidant decay levels, as well as calculations to prevent sediment from settling.

The tanks would be fitted with pressure-vacuum relief valves that only open when the ballast water is being transferred or occasionally to relieve built-up pressure or vacuum from a diurnal cycle. The gauging system would be a closed system to limit contaminants from entering the tanks. At least two tank vents would be installed. Each vent would be fitted for ready connection to ventilation blowers to facilitate gas-freeing tanks to make safe for personnel entry.

Depending on the required oxidant level, the ballast tanks might also require a special coating system. In addition, piping system gaskets and valve seals might also require special materials not typically used in seawater applications.

Discharge

Each tank would be fitted with piping for deballasting with a high suction at approximately 300 mm above the tank bottom, and a low suction at approximately 75 mm above the tank bottom. The high suction would be used for ballast tank discharge, such that the discharge does not contain sediment. The low suction would be used for stripping sediment from tanks when suitable disposal facilities are available.

The discharge piping would be independent from the uptake piping. Each tank would be outfitted with an isolation valve connecting it to the discharge main header. The header would lead to a reactor tank of one-meter height built into the ship's double bottom with at least 25 m^3 capacity, allowing a contact time of at least three minutes. During the contact time, the oxidant level would be neutralized and water quality confirmed prior to discharge. The system would be failsafe, returning the ballast water to the ballast water storage tank if needed.

A dedicated seawater overboard, designed to avoid contamination from ballast water uptake or other sources, would be fitted for discharging the ballast water. The disinfection step would be as close as practical to the overboard. This final disinfection step would provide assurance against contaminants in the reactor tank where the oxidant was neutralized, as well as providing a measure of caution in treating the ballast water a second time by a different process.

The ballast water could be moved through the discharge by gravity if there is adequate head in the ballast tank. At any time, a pump would take suction on the reactor tank, avoiding pump contact with the oxidants. The pump would then discharge to the UV unit and overboard.

Summary

In summary, reaching the USCG Phase-2 standard, or even 100 times the IMO D-2/ Phase 1 standard, will likely require wholly new treatment systems. Such new systems will have many attributes different from existing BWMS. They will use new technological devices, including those drawn from the water treatment industry; employ multistage treatment processes; emphasize technological process controls and multiple monitoring points to ensure desired performance, rather than rely on end-of-pipe testing; include physical barriers to minimize the potential for cross-contamination of the system; and become part of an integrated ballast water management effort. These new approaches will achieve higher performance, but will require time to develop, test, and determine their practicality and cost.

In addition, new BWMS technologies will need to become more energy-efficient. Driving factors include rising fuel costs, potential future valuations or other constraints on air emissions and other pollutants, and potential future taxes of carbon sources from maritime shipping. To date, attempts to meet proposed discharge standards generally have increased the energy required for ballast management. New BWMS methods should attempt to reverse this trend. Recent innovations have significantly reduced the volume of discharged ballast water, and in some cases eliminated discharges in all routine operations. Such direct approaches should continue and their reduced environmental impact should be recognized and encouraged in regulatory, monitoring and enforcement efforts. These approaches are discussed further in response to charge question 4.

5.4. Recommendations for Addressing Impediments and Constraints

Charge question 3c. What recommendations does the SAB have for addressing these impediments and constraints?

Several existing technologies have demonstrated compliance with the D-2 standard during testing periods. However, it is not clear that these BWMS will operate consistently at this level of performance on board the many thousands of vessels that will require their use. Clearly defined and transparent programs for compliance monitoring and enforcement are needed to promote consistent, reliable operation of BWMS; such programs do not yet exist. Ideally, vessel crew members would have the technological capability to monitor BWMS efficacy, and make real-time corrections to maintain compliance. Further, it is important that BWMS manufacturers document and report performance metrics beyond discharge treatment efficacy. This information would enable vessel operators to select systems that best integrate with their operations. For example, the ETV protocol provides guidance for third-party evaluation of factors such as energy consumption and reliability. These and similar metrics should be encouraged.

Although meeting significantly higher standards will likely require completely new treatment approaches, the Panel can predict neither which combination of treatment processes will achieve the highest efficacy nor their ultimate performance. The Panel recommends that one or more pilot projects be commissioned to explore new approaches to ballast water treatment, including tests of ballast water transfer and treatment at a reception facility

5.5. Impediments Based on Organism Type

Charge Question 3d. Are these impediments more significant for certain size classes or types of organisms (e.g., zooplankton versus viruses)?

Shipboard impediments apply to all size classes of organisms and specified microbes. This broad conclusion is based on analysis of BWMS test results, as well as general considerations of the treatment processes and the vessel application constraints.

With regard to specific technologies, however, BWMS performance varies across target organisms. For example, existing BWMS are capable of removing (e.g., mechanical separation) or killing (e.g., deoxygenation, UV, chlorine dioxide) the great majority and in some cases, nearly all organisms ≥ 50 μm; UV irradiation kills or inactivates unicellular organisms and viruses more efficiently than it does metazoans; and deoxygenation does not eliminate bacteria but rather alters microbial communities.

Such variation among organisms is exemplified by testing data. Section 4 of this report reviews results of seven BWMS that "reliably met" the IMO D-2 standard. All treatment systems were limited in their ability to reach extremely stringent, proposed standards for total bacteria. In addition, although they met IMO D-2 standards, some live organisms were found in either one or both of the ≥ 10 to < 50 μm and ≥ 50 μm size classes. In summary, these data indicate that current technology is broadly challenged by bacterial counts and sometimes selectively challenged by both ≥ 10 to < 50 μm and ≥ 50 μm size classes.

5.6. Sterilization of Ballast Water Discharge

Charge Question 3e. Can currently available treatment processes reliably achieve sterilization (no living organisms or viable viruses) of ballast water onboard vessels or, at a minimum, achieve zero or near zero discharge for certain organism size classes or types?

It is an unrealistic and unattainable goal for current BWMS to yield ballast discharge that is "sterile", i.e., "free from living organisms and viruses" (Madigan and Martinko 2006). Given the volumes of water requiring treatment, sterilization is not possible using current technologies; there simply is not enough energy on a vessel to implement steam autoclaving of its ballast water tanks and piping systems. With respect to "zero or near zero discharge", however, technology exists to remove most organisms in the size classes ≥ 10 to < 50 μm and ≥ 50 μm. As a practical matter, the Panel notes that it is not possible to measure zero (sterilization) or near zero discharge—especially for microorganisms such as phytoplankton, bacteria, and viruses, which are especially difficult to differentiate as "live" or "dead" on the basis of physiological certainties. If such values cannot be measured, a BWMS cannot be controlled to ensure zero or near-zero discharge at the "end of pipe" for a working vessel.

6. Limitations of Existing Studies and Reports

6.1. EPA's Charge Question

This section responds to Charge Question 4: "What are the principal limitations of the available studies and reports on the status of ballast water treatment technology and system performance and how can these limitations be overcome or corrected in future assessments of the availability of technology for treating ballast water onboard vessels?" Bearing in mind the broader charge to "provide advice on technologies and systems to minimize the impacts of invasive species in vessel ballast water discharge" (Feb. 2010 Federal Register notice), this section addresses aspects of ballast water discharge not covered in the responses to earlier charge questions. Several themes emerged which the Panel discusses in the following sections. First, improved methods for testing and reporting are needed to ensure that high quality data are available with which to assess BWMS performance. Second, improved data also are important to the development of effective approaches to enforcement and compliance. Third, existing data and reports on the effectiveness of practices and technologies other than shipboard BWMS are inadequate because insufficient attention has been given to integrated practices and technologies that could reduce the risk of invasions. These include managing ballast water uptake to reduce presence of invasive species, reducing invasion risk through operational adjustments and changes in ship design to reduce or eliminate the need for ballast water, and consideration of land-based reception facilities for ballast water treatment. Voyage-based risk assessments could be used to integrate such practices, through use of applied risk management principles such as Hazard Assessment and Critical Control Points methods.

6.2. Testing Shipboard Treatment Systems: Protocols, Analysis, and Reporting Practices that Could be Improved

This section applies to test facilities both in the U.S. and abroad and was informed by the ETV Protocol for land-based verification of BWMS performance (U.S. EPA 2010, hereafter the Protocol). The Panel acknowledges the many efforts put forth by various technical panels and stakeholder groups over many years to draft, validate and finalize the Protocol. Most of this section focuses on land-based verification testing (used to gain Type Approval from Maritime Administrations) rather than shipboard verification testing (also used to gain Type Approval) or compliance testing (used to determine adherence to any discharge standard when a vessel enters a port of call). This is because, to date, programs that address these types of testing have not been finalized in the U.S.

6.2.1. Confusion of Research and Development and Certification Testing

In some cases, little if any distinction is made between research and development (R&D) testing and verification testing. Adjustments to BWMS often are made during testing of prototypes and, in some cases, only the most favorable results are reported. Thus, certification may be gained on the basis of unrealistically favorable results that may not be representative of replicated testing with multiple commercially available units of a BWMS. To address this problem, the Protocol requires that BWMS undergoing verification testing are "prefabricated, commercial-ready treatment systems" and that all test results be reported (U.S. EPA 2010). Given the early state of the BWMS industry, mass-produced assembly line systems are not currently tested. As indicated in the Protocol, R&D testing should be barred from use in certification testing.

To ensure that the performance of ballast water treatment systems is objectively and thoroughly evaluated during verification testing, experienced specialists in an independent testing organization should conduct the tests (as required in the Protocol), rather than the system manufacturers. This is important because research has shown that it is extremely difficult for system creators—who have constructively designed their systems—to change their perspective and instead view their system from the "deconstructive" state of mind that is focused on finding flaws and exposing weaknesses and limitations (Myers 1979). Thus, it is critically important that verification testing be conducted by independent specialists in order to assess system performance in a scientifically rigorous way. Further, as noted in the Protocol, the credentials of these personnel should be approved by the Verification Organization (the entity that oversees testing preparation, testing and the Verification Report issued by the test facility at the conclusion of testing). In sum, testing should be conducted by a party that is independent from the manufacturer and has appropriate, established credentials.

6.2.2. Lack of Standardized Testing Protocols

Comparative evaluations of the performance of different BWMS are hampered by inconsistencies in discharge standards and in testing protocols. As shown in Table 2-1, there are diverse state, national and international discharge standards for ballast water—including differences in limits that vary by orders of magnitude for similar categories of organisms. This range of standards not only results in confusion for the regulated industry but also provides significant challenges for testing of BWMS. Performance standards set requirements for technology to achieve and should help to advance progress in treatment system designs, but only if a set of standardized, practical, scientifically rigorous assessment techniques is available to evaluate system performance. The IMO standards are based upon different size groups of organisms, and all size groupings pose challenges for assessing performance.

Comparison of the performance of different ballast water treatment technologies requires consistent testing protocols regardless of the target discharge standard (Phillips 2006; Ruiz et al. 2006). To date, all BWMS have been evaluated using the basic approaches provided by the IMO Guidelines for Approval of Ballast Water Management Systems (G8) and the Procedure for Approval of Ballast Water Systems that Make Use of Active Substances (G9) (IMO 2008a,b). While G8 and G9 suggest a basic framework, the level of detail required for rigorous and comparable BWMS testing is lacking. The state of California also has developed ballast water treatment technology testing guidelines that are intended to provide a standardized approach for evaluating treatment system performance (Dobroski et al. 2009). Procedures also are are being developed for verifying vessel compliance with California performance standards.

The 2010 Protocol is a federal program that is much more detailed and proscriptive regarding test facility design, sampling design and volume, sample handling, analytical methods, data reporting and QA/QC requirements. However, this Protocol has yet to be implemented in practice or broadly adopted. Thus, at present there is no broad international program that includes performance standards, guidelines, and protocols to verify treatment technology performance, and no standardized sets of methods for sampling and analysis of ballast water to assess compliance. The existing federal and various state standards lack consistency as well. Treatment evaluations generally are designed to test whether a given technology can meet IMO D-2 standards in accordance with the IMO G8 or G9 Guidelines (IMO 2008 a, b).

With exception of BWMS installed aboard vessels enrolled in the U.S. Coast Guard's (USCG) Shipboard Technology Evaluation Program (STEP), BWMS presently are not approved for use in compliance with proposed federal ballast water management requirements. Thus, while there are various state ballast water management requirements, there is no formal U.S. type approval program for BWMS (although one is described in the USCG proposed final rule). The EPA has, however, included provisions in the draft NPDES Vessel General Permit for vessels with treatment systems that discharge ballast water containing biocides or chemical residues.

Performance standards set requirements for technology to achieve and should help to advance progress in treatment system designs, but only if a set of standardized, practical, scientifically rigorous assessment techniques is available for use in assessments. All existing and proposed performance standards are based upon different size groups of organisms, and all size groupings pose challenges for accurate assessments of treatment performance (see below). In the IMO D-2 performance standard, organisms in the < 10 μm size class are represented by a subset of taxa consisting of three indicator bacteria or bacteria groups (*Vibrio cholerae, Escherichia coli*, and Enterococci). Assessment has relied upon a subset group of organisms as representative of treatment of all bacteria (see Section 6.2.3, below). There is as yet no strong evidence for suitable proxy organisms to represent the virus size class, and no acceptable methods to verify compliance with a total virus standard (which is in the proposed USCG Phase 2 performance standard).

The following section provides Panel recommendations for future versions of the ETV Protocol, thus it focus on differences from the Protocol rather than reiterating recommendations made in the Protocol. Because the Protocol pertains only to land-based verification testing (not shipboard testing), the Panel's recommendations focus on land-based testing for verifying treatment performance by independent testing operations. This section also comments on shipboard testing.

Test Verification Factors

The Protocol recommends that all treatment systems be verified using the following factors: biological treatment efficacy (or BE, defined as the removal, inactivation or death of organisms), operation and maintenance (O&M), reliability as measured by the mean time between failures (MTBF), and environmental acceptability including residual toxicity and safety. The Panel agrees with the Protocol that biological treatment efficacy should be measured as the concentration, in the treated ballast water discharge, of the organism size classes indicated in the IMO D-2 and USCG Phase 1 performance standards, with a minimum concentration of organisms in the control tank discharge. Other measurements can include water quality parameters in comparison to appropriate water quality standards. Verification protocols should include detailed descriptions of on-site sampling, sample handling (chain of custody), QA/QC, in-place mechanisms for selecting independent laboratories with appropriate expertise and certification to conduct the sample analyses, and requirements for compliance reporting.

The Panel also agrees with the Protocol that tests and species selected for toxicity testing during commissioning need to have carefully justified protocols detailed in the Test Plan. BWMS that involve a chemical mode of action are regulated under the National Pollutant Discharge Elimination System (NPDES) permit process (Albert et al. 2010), which requires demonstration of "no adverse effects" as evaluated through chemical-specific parameters and standardized Whole Effluent Toxicity (WET) testing (U.S. EPA 2002a-c; 40 CFR 136.3, Table 1A). WET experiments are designed to assess the effects of any residual toxicity on beneficial organisms in receiving waters. Standardized acute and chronic toxicity assays have been developed by the EPA for a limited number of freshwater and marine species (Table 6-1). The Protocol does not include specific freshwater assays, but recommends that toxicity tests for biocide treatments in brackish and marine waters should be selected from the EPA acute toxicity assay for mysids (EPA OPPTS Method 850.1035; http://www.epa.gov/opptsfrs /OPPTS_Harmonized/850_Ecological_ Effects_Test_Guidelines/Drafts/850-1035.pdf), and the chronic toxicity assays for the inland silverside *Menidia beryllina* (larval survival and growth, EPA Method 1006.0; http://www.epa.gov /OST/WET/disk1/ ctm13.pdf) and the sea urchin, *Arbacia punctulata* (fertilization, EPA Method 1008.0; http://www.epa.gov/OST/ WET/disk1/ctm15.pdf). The Panel recommends that freshwater assays also be included in toxicity testing.

The Protocol also recommends that complete results of verification testing, including equipment failures, be reported as standard practice. These data are needed to enable realistic evaluation of a given BWMS. At present, there is no requirement under the IMO G8 guidelines to report tests in which a BWMS does not perform to the D-2 performance standard. The Panel strongly recommends that reports should include all test results, and that criteria for approval should consider the failure rate (proportion of tests that were successful).

Challenge Conditions

The Panel recommends that testing should be applied across the gradient of environmental conditions (temperatures, salinities) represented by the Earth's ports; to address this concern, the ETV Protocol requires testing at a minimum of two salinities (U.S. EPA 2010), although some Panel members argued the minimum should be three. All treatment technologies should function well across the range of physical/chemical conditions and densities/types of biological organisms that a ship encounters. Thus, BWMS ideally should be verified using a set of standard challenge conditions that encompass the suite of water quality conditions, and that capture environmental conditions represented by ports and a range of densities of the organisms and organism size classes (unless a BWMS is designed, and certified, for only a specific subset of conditions).

Table 6-1. Freshwater and marine species for which the U.S. EPA has developed standardized acute and chronic toxicity assays (http://www.epa.gov/waterscience/WET)

Habitat	Acute Toxicity	Chronic Toxicity
Freshwater		
Algae	---	*Selenastrum capricornutum* (growth)
Zooplankton	*Ceriodaphnia dubia* *Daphnia magna* *Daphnia puplex*	Survival, reproduction --- ---
Fish	Bannerfin shiner (*Cyprinellale edsi*) Brook trout (*Salvelinus fontinalis*) Fathead minnow (*Pimephale spromelas*) Rainbow trout (*Oncorhynchus mykiss*)	--- --- Larval survival, growth; embryo-larval survival, teratogenicity ---
Marine		
Mysid shrimp	*Americamysis bahia*	Survival, growth, fecundity
Sea urchin	---	*Arbacia punctulata* - fertilization
Fish	Sheepshead minnow (*Cyprinodon variegatus*) Silversides (*Menidia beryllina, M. menidia, M. peninsulae*)	Larval survival, growth; embryo-larval survival, teratogenicity *M. beryllina* - larval survival, growth

Table 6-2. Comparison of the ETV Protocol's recommendations (U.S. EPA 2010) and the alternatives the Panel recommends that EPA consider, with respect to minimum criteria for challenge water total living populations, criteria for a valid biological efficiency (BE) test cycle at land-based facilities (living organisms in control tank discharge after a holding time of at least 1 day), and water types (salinity groupings) for completion of BE tests.[1] Three salinity ranges are recommended for BWMS that are planned for use in freshwater, brackish, and marine waters

Minimum Criteria for Challenge Water Total Living Populations; and Criteria for a Valid BE Test Cycle - Living Organisms in Control Tank Discharge After 1 Day Holding Time		
Size Category[2]	**ETV Protocol**	**Panel Alternatives**
> 50 μm	105 organisms m-3, 5 species in 3 phyla	same
> 10 μm and < 50 μm	103 organisms mL^{-1}, 5 species in 3 phyla	same
Other[3]	< 10 μm: 103 mL^{-1} as culturable aerobic heterotrophic bacteria	< 10 μm: 10^3 selected protists mL^{-1} < 2 μm: same as ETV for < 10 μm
Water Types (Salinity Groupings) for Completion of BE Tests[3]		
Fresh (salinity < 1)	Two salinity ranges;	Two or three salinity ranges;
Brackish	brackish ≡ salinity 10-20;	brackish ≡ salinity 1 to < 28
Marine	marine ≡ salinity 28-36	marine ≡ salinity > 28
Physical/Chemical[4]		
Environmental:	Temperature (4-35°C), DOC, POC, TSS, MM, pH, DO	
Others of Specific Interest:	Example - nutrient concentrations (TN, TP, TKN, NH_x, NO_x, SRP)	

[1] Abbreviations: DOC, dissolved organic carbon; POC, particulate organic carbon; TSS, total suspended solids = particulate organic matter (POM) + MM (mineral matter); DO, dissolved oxygen; TN, total nitrogen; TP, total phosphorus; TKN, total Kjeldahl nitrogen; NHx, ammonia + ammonium; NOx, nitrate + nitrite; SRP, soluble reactive phosphorus.

[2] Size ≡ maximum dimension on the smallest axis.

[3] Effects on culturable aerobic heterotrophic bacteria are assumed to be indicative of effects on all bacteria.

[4] The ETV Protocol's water quality challenge matrix for verification testing includes the following minimum water characteristics for the three salinity water types as: Dissolved organic matter, 6 mg L^{-1} as DOC, and particulate organic matter, 4 mg L^{-1} as POC; MM 20 mg L^{-1} and TSS 24 mg L^{-1}; and temperature range 4-35°C.

The ETV Protocol states that the objectives for challenge conditions are to verify treatment system performance using a set of "challenging, but not rare, water quality conditions representative of the natural environment," and to verify removal or kill of organisms ranging in size from bacteria to zooplankton, using natural assemblages and appropriate analytical techniques that enable quantification of densities of live organisms (U.S. EPA 2010, p.18). It is important to evaluate the effectiveness of treatment systems under conditions that challenge the technology because certain water quality conditions can interfere with some treatment processes. These physical/chemical environmental conditions are generally understood and relatively few in number, which helps to limit the number of water quality metrics that must be included in the protocol (Table 6-2).

In recognition of the difficulties that can be encountered, especially in ship-based testing, tests of the three salinity ranges could include two land-based tests and one ship-based test. A rationale for recommending tests of all three salinity ranges is that if a given BWMS is planned for use across all three salinity ranges, but testing indicates that its efficiency at organism removal is poor under one or more of the salinity groupings, then that system should not be used by ships visiting ports that are characterized by such conditions. Similarly, if a BWMS is planned for use across other environmental gradients (e.g., temperatures from cold to warm waters, or salinities from fresh to marine), but tests indicate that it has poor efficiency in removing biota under part of the natural range, then that system should not be used by ships visiting ports that have such conditions. Indeed, the USCG proposed rule indicates that "at least 2 sets of test cycles should be conducted with different salinity ranges and associated dissolved and particulate content as described. BWMS not tested for each of the 3 salinity ranges and water conditions listed in this section may be subject to operational restrictions within a certificate of approval" (USCG 2009, p. 44666). A fully crossed design should be used where possible, for example, if natural water can be obtained at the desired salinity range. As another example, cold water testing may be critical to understand the breakdown of chemical treatments (i.e., active substances), but testing with natural water in the winter would encounter relatively few organisms in the challenge water, making it difficult to achieve recommended challenge conditions.

There are other major practical constraints on such tests. First, alterations to establish the natural range of physical and chemical conditions should be imposed without affecting the concentrations, diversity and viability of the biota present. For that reason, natural water sources ideally should be used to impose the levels of salinity, rather than artificially modified salinity. Artifactual interactions may occur between biota and artificial media (e.g., artificial seawater prepared with commercially available "sea salts"). The Panel thus diverges from Anderson et al. (2008) in recommending that a source of filtered, high-quality natural freshwater or seawater should be used to prepare treatments insofar as possible. There are pros and cons with either approach: Artificial sea salts are expensive but enable routine preparation of media. However, caution is warranted in using artificial sea salts because some ingredients that are not found in natural seawater, such as phthalate esters (e.g., di(2-ethylhexyl)phthalate, a commonly used plasticizer in Instant Ocean aquarium salts), are abundant and can be toxic to aquatic life, resulting in spurious data (e.g., Peal 1975; Moeller et al. 2001).

In addition, various dissolved organic compounds that are important to the nutrition and the life histories of aquatic organisms likely will be missing from artificially constructed media. While use of natural waters avoids such problems, the natural water source should be as free as possible from toxic pollutants, which are increasingly ubiquitous in fresh, brackish, and coastal marine waters (Kay 1985; Pate et al. 1992; Loganathan and Kannan 1994; Hoff et al. 1996; U.S. EPA 2000; Shaw and Kurunthachalam 2009), or contain at most only trace levels of such pollutants. Final selection of natural versus an artificial water sources requires careful consideration of these issues. In addition, when using artificial water sources or otherwise modifying environmental conditions, timing is important; care should be taken to avoid imposing rapid environmental changes that, alone, could stress or kill the biota tested.

Similarly, the Protocol recommends adjusting particulate organic matter (POM), if natural waters do not meet challenge conditions, by adding commercially available humic materials, plankton, detritus, or ground seaweed; commercially available clays can be added to adjust the mineral matter concentration (U.S. EPA 2010). However, the Panel is concerned that the cation exchange capacity of the dried, then rehydrated, clays can significantly alter plankton communities (Avnimelech et al. 1982; Burkholder 1992; Cuker and Hudson 1992). Artificial modification of dissolved organic carbon (DOC) is difficult to achieve without a strong potential of affecting the biota present, especially the smaller size-fraction components. The Panel believes that the testing organization should be required to verify, insofar as possible, that in preparing the test water, any materials added had minimal effects on the biota, and "minimal effects" should be clearly defined.

The IMO (2008a, b), the Protocol, and other suggested standards (e.g., California VGP 401 certification/State regulations; see Albert et al. 2010) make no mention of protists in the < 10 μm size range. Many harmful organisms occur in this size range (e.g., harmful "brown tide" pelagophytes *Aureococcus anophagefferens* and *Aureoumbra lagunensis*, many harmful cyanobacteria, and certain potentially toxic dinoflagellates; see Burkholder 1998, 2009). The selected bacteria presently targeted for standards are not useful as indicators for these taxa which, as a general grouping, can adversely affect both environmental and human health (Burkholder 1998, 2009). Thus, failure to consider this size class represents a serious omission in efforts to protect U.S. estuaries, marine waters and the Great Lakes from harmful species introductions. For some of these taxa, such as toxigenic *Microcystis* spp. affecting some of the Great Lakes (e.g., Boyer 2007), the tendency of the cells to aggregate into colonies can sometimes "boost" them into the > 10 μm size range during filtration processes, but the size measurements are based on individual cells.

There is a critical need to consider harmful representative protists (which should be expected to vary depending on the geographic region) from this size class in developing protective ballast water standards. Depending on the salinity and the region, and based on the smallest cell dimension, examples of candidates could include selected toxigenic cyanobacteria such as *Anabaena flos-aquae*, the haptophyte *Prymnesium parvum*, brown tide organisms *Aureococcus anaphogefferens* or *Aureoumbra lagunensis*, small toxigenic dinoflagellates such as *Karlodinium veneficum*, and the pathogenic protozoans *Giardia* spp. and *Cryptosporidium parvum* that are found across the salinity gradient. Accordingly, protists in this size class should be included in standards for assessing the performance of BWMS in land-based testing if they were naturally occurring at the test facility. For shipboard verification testing or compliance testing, where the source water is unknown until sampling occurs, appropriate organisms for evaluation can be selected accordingly.

Verification Testing

The Panel offers considerations that differ from the Protocol on some points, including specifics for collecting water quality and biological samples for verification testing of BWMS (Table 6-3). Some panel members argue that these points should be a part of verification testing; others argue that such approaches could be incorporated into future revisions of the Protocol if their utility, effectiveness, and practicability were deemed appropriate. Some Panel members have concerns that such analyses would (1) provide qualitative indications of viability, not the quantitative data on the density of organisms necessary to assess BWMS performance in accordance with a discharge standard, or (2) overestimate the number of viable organisms. Also, conducting new analyses, in addition to those required in the Protocol, might not be practicable by already busy testing teams.

For zooplankton, phytoplankton and other protists, the Panel supports the need for collecting at least 3-6 m^3 of sample volume at each required location on a time-averaged basis over the testing period. Field quality control samples and field blanks should be taken under actual field conditions to provide information on the potential for bias from problems with sample collection, processing, shipping and analysis (Ruiz et al. 2006). Accepted scientific methods should be used for all analyses (e.g., for water quality parameters, U.S. EPA 1993, 1997; American Public Health Association (APHA) et al. 2008). Biological samples should be collected in a time-integrated manner during the tests, and sample collection tanks should be thoroughly mixed prior to sampling to ensure homogeneity (U.S. EPA 2010). Samples collected from control and treated tank discharges should be taken upstream from pumps or other apparatus that could cause mortality or other alterations, and if pumps and valves must be used upstream of sample collection, they should previously have been tested and shown not to damage organisms (U.S. EPA 2010). Note that analysis of some parameters is extremely time-sensitive (Table 6-3). For example, zooplankton die-off occurs in some samples held for 6 hours or more (U.S. EPA 2010); the timing of die-off likely varies depending on the zooplankton community, and the upper limit should be determined at each test facility. The approximate maximum hold times should maintain detectable zooplankton mortality over time at < 5%.

An alternative to quantifying viable organisms from unpreserved samples is preserving samples immediately upon collection and then assessing intact organisms as "viable when collected." Some Panel members recommend this as a more practical alternative, whereas others note vigorous validation is needed before it can be recommended. This approach is commonly used in characterizations of microflora and microfauna assemblages in the peer-reviewed literature, based on the fact that protists and zooplankton deteriorate quickly once dead (within minutes to hours: Wetzel 2001; Johnson and Allen 2005; and see Section 6.2.3). Effective "fast-kill" preservatives can be used that cause death before distortion or cell lysis can occur. Standardized, accepted techniques are available for quantifying "viable when collected" protists and zooplankton from preserved material (e.g., Lund et al. 1958; Wetzel and Likens 2000; Johnson and Allen 2005; and see Section 6.2.3). A shortcoming of this approach is that dying organisms which still contain apparently intact cellular contents would be included in the "viable" estimate. Thus, the number of viable organisms would be overestimated if a large fraction of the sample was dead. In addition, as for counts based on unpreserved material, it is difficult to assess whether some resistant structures such as thick, opaque cysts contain organisms with intact cell contents. Because of practical and environmental health/safety constraints, neither approach avoids the problem of likely major losses of viable organisms that occur during rapid concentration of large sample volumes. If a "formerly viable" approach were used, the Panel recognizes that it would need to be validated and approved by the Verification Organization.

6.2.3. Compromises Necessary Because of Practical Constraints in Sampling and Available Methods

Ideally, the goal of standard challenge conditions would include the full range of (a) challenging conditions present in the world's ports, (b) organism density, (c) taxonomic diversity, and (d) organism size classes. Meeting this ideal goal is impeded by several serious practical constraints in sampling large ballast tanks effectively, and in the methods that presently are available for quantifying viable organisms. As Lee et al. (2010, p.19) pointed out, "perfect compliance and no failure is practically, if not theoretically, impossible, particularly for microbiological organisms unless ballast water is discharged into a land-based treatment facility or ships are redesigned to eliminate the need to discharge ballast water." This section considers how the ideal can be modified to accommodate practical considerations while accomplishing a meaningful evaluation of the efficacy of BWMS.

Table 6-3. Sample volumes, containers, and processing for core parameters and auxiliary nutrients (nitrogen, N; phosphorus, P; silicate, Si; carbon, C). Note that HDPE ≡ high-density polyethylene, and POC information is from Baldino (1995). Recommendations that differ from those in the ETV (U.S. EPA 2010) are indicated in bold[1]

Parameter	Minimum Sample Volume	Containers	Processing/Preservation Holding Time	Maximum
TSS	100 mL	HDPE or glass	Process immediately or store at 4oC	1 week
DOC	25 mL	glass	Pre-combusted GF/F filters; **preserve filtrate with H_3PO_4 (pH < 2), hold at 4° in darkness (APHA et al. 2008)**	28 days
POC	500 mL	HDPE	Filter (GF/F in foil); freeze filter	28 days until analysis
MM	= TSS -POC	----	----	----
DO	300 mL **or *in situ* sensor**	glass BOD bottles	Fix (Oudot et al. 1988); titrate in 2-24 hours; or Continuously recording	24 hours
Chlorophyll *a*,[1] pheopigments	400 mL	dark HDPE	Filter (GF/F); fix with saturated $MgCO_3$ solution; freeze filter until analysis	3 weeks
Phytoplankton No.[2] (viable, < 10 µm - selected harmful taxa)	**500 mL**	**dark HDPE**	**Filter (Nuclepore or Anotech); assess autofluorescence (e.g. MacIsaac and Stockner 1993), *or* Filter, fix (e.g. 0.2% (v/v) formalin), freeze filter; *or* filter, fix, followed by selected molecular techniques (e.g. Karlson et al. 2010)**	**Process immediately 3-4 weeks months**
Phytoplankton No. (viable, nano-/micro-plankton, > 10 to 50 µm **and > 50 µm**)[3]	3 m^3 (1,000 L) → 1 L	60 mL dark HDPE	Viable: No preservative; stain with FDA, CMFDA; ***or*,** **fix with acidic Lugol's solution (Vollenweider 1974), store at 4°C in darkness, and quantify as viable when collected *and* combine with various molecular techniques to confirm harmful taxa of interest (e.g. Karlson et al. 2010)**	process immediately; **or 28 days, preferably 1 week**

Table 6-3. (Continued)

Parameter	Minimum Sample Volume	Containers	Processing/Preservation Holding Time	Maximum
Other protists (#)[1] (viable heterotrophs, < 10 μm – selected harmful taxa)	**500 mL**	**100 mL, dark HDPE**	**Techniques appropriate for the selected taxa (e.g. U.S. EPA 2005)**	**variable**
Zooplankton # (viable, > 50 μm)	3 m^3 (3,000 L)[4] → 1 L	1-L flask	No preservative; subsample 450 1-mL wells3 and probe; **fix with buffered formalin and Rose Bengal's solution to quantify; *or***	Process immediately (< 6 hr) [5]
Zooplankton # (viable) (cont'd.)			**fix as above and quantify as formerly viable (Johnson and Allen 2005)**	**Process within 1 month**
Bacteria (active culturable, aerobic heterotrophic - selected taxa)	≥1000 mL	sterile HDPE	Plate on appropriate media (U.S. EPA 2010)[6]	Process Immediately
Nutrients[1] -				
TN, TP total Kjeldahl N (TKN)	60 mL	varies	**Varies – see standard methods (U.S. EPA 1993, 1997; APHA et al. 2008; and U.S. EPA 2010, p.39)**	varies (mostly 28 days)
NO_xN, NH_xN, SRP, **SiO2**	60 mL	varies	Varies – see standard methods as above	varies (mostly 28 days)

[1] Methods in the ETV Protocol that differ from those and recommended above by the Panel for consideration are as follows: DOC – pass sample through a GF/F filter and freeze filtrate until analysis; chlorophyll *a* and pheopigments – listed as an auxiliary parameter rather than a core parameter; protists (phytoplankton, protozoans) in the < 10 μm size class are not considered; TN, TP, total Kjeldahl N, and silica are not addressed; and dissolved inorganic phosphate is referred to here as soluble reactive phosphorus (SRP).

In situ sensors are available for measuring chlorophyll *a* as relative fluorescence units, but not as chlorophyll *a* concentrations. Chlorophyll *a* may be considered as a core parameter or as an auxiliary parameter, used as a collective indicator for algal biomass. The Panel also recommends assessment of nutrients (TN, TP, total Kjeldahl N, and silica) if possible, although nutrients are not considered as core parameters by the ETV and the Panel recognizes that core parameters should have top priority.

The Panel also recognizes the fact that present performance standards are for living (viable) organisms. Because of widely acknowledged practical limitations in techniques to assess living (viable) organisms, in this table and explained in the writing below, the Panel suggests alternate techniques for quantifying the organisms that were viable in samples when collected.

[2] This size category has not been considered for ballast water treatment standards by IMO (2008a,b), the ETV (U.S. EPA 2010), etc. Because many harmful organisms occur in the < 10 μm grouping, this size class should be considered for inclusion in assessment of BWMS.

[3] FDA, fluorescein diacetate; CMFDA, 5-chloromethylfluorescein. Delicate protists (e.g. wall-less flagellates) mostly would not be expected to survive the process of rapid concentration of large-volume samples. The Panel recognizes that samples collected for protists are not concentrated in the same way as samples collected for the larger size class: For the P-1 standard, 3-6 L (taken as whole water, isokinetically from the discharge of control or treatment tanks) are concentrated to 1 L, which can be done with a sieve. Nevertheless, this process can lyse delicate protists. As a more practical alternative than attempting to quantify viable algae and other protists from unpreserved samples, an option is to preserve samples immediately upon collection and then assessing intact organisms as "viable when collected," based on the fact that protists such as most algae in this general size class are known to lyse and/or decompose rapidly (minutes to several hours) after death, so that the cell contents become distorted or are lost even if the cell coverings remain (Wetzel 2001). It should be noted that vital (or mortal) stains address the question, "Is this alga living?" in a way that is substantially different than the presence/absence of intact chloroplasts and other cell contents. Thus, the two methods sometimes do not yield the same quantitative answer and require careful calibration.

[4] It should be noted that phototrophic organisms in this size class should be quantified using the protocols for phytoplankton, above.

[5] Zooplankton die-off occurs in samples held for 6 hours or more (Naval Research Laboratory, unpubl. data; U.S. EPA 2010).

[6] Media suggested by the U.S. ETV (2010, p.47) for brackish/marine taxa include 2216 Marine Agar and salt-modified R2A agar; media for freshwater species may include Plate Count Agar and Nutrient broth (plus agar (15 g L^{-1}).

Standardization of Choices of Standard Test Organisms

The Protocol defines standard test organisms (STOs) as "organisms of known types and abundance that have been previously evaluated for their level of resistance to physical and/or chemical stressors representing ballast water technology" that are used in bench-scale testing (U.S. EPA 2010, p. 74). The Protocol (U.S. EPA 2010) requires that prior to full-scale verification testing, laboratory experiments be conducted to evaluate post-treatment viability of STO taxa used to assess the biological effectiveness of BWMS in removing zooplankton, protists (heterotrophic and phototrophic), and bacteria.

The selection and development of STOs that are broadly resistant to treatments for use in testing BWMS performance is a fertile field of research because of the practical need (Hunt et al. 2005; Anderson et al. 2008; U.S. EPA 2010). The Panel urges caution in the use of STOs, however, since results from a very small number of taxa are broadly applied to all of the organisms in the same general grouping (e.g., protozoans in a certain size class). An assumption that first must be validated is that the selected taxa are among the most resistant to treatment, so that most organisms are eliminated when the surrogate taxa are eliminated (Ruiz et al. 2006). The fundamental challenge is to identify the best species that are "representative" of a broad range of organisms within a given size class. Good candidates are considered to be easily and economically cultured in large numbers for future full-scale testing in experimental ballast water tanks; tolerant of a wide range of environmental conditions; reliable and consistent in their response to treatment across culture batches; and resilient in withstanding ballast water tests and sampling (Ruiz et al. 2006; Anderson et al. 2008). A list of suggested STOs is provided in the ETV Protocol. An obvious risk is spurious results from surrogate taxa that poorly represent the larger group of organisms in a given size class. Because testing with STOs is part of a larger testing process (land- based and shipboard verification testing) that employs a range of organisms, this risk is somewhat ameliorated.

Protocols for STOs should include clear justification for use of these taxa under a defined set of conditions; careful consideration of potential confounding interactions between the STOs and natural species; and the percentage ratio of challenge organisms that are STOs versus naturally occurring taxa in the challenge water. Selection of a specific combination of STOs should be based upon extensive testing at bench and mesocosm scales, preferably by several laboratories located in different geographic regions, of a wide range of STO species, life histories, habitats, and source regions across environmental gradients (Ruiz et al. 2006). Consistent use of the same protocols is needed in order to minimize confounding factors and strengthen comparability. Ideally, several STOs or taxonomic subgroups, including several life stages, should be included in the tests since confidence in interpretations can be strengthened by this redundancy. It would also be best to include multiple strains (populations) of candidate STOs if possible, to account for significant intraspecific variability in response to environmental conditions that is commonly documented, particularly among protists (Ruiz et al. 2006; Burkholder and Gilbert 2006). Nevertheless, the Panel recognizes that practical and economic considerations may prevent testing beyond use of one or two STOs to represent each size class.

Standardization of Choices of Indirect Metrics (Surrogate Parameters)

There are practical and logistical limitations involved in obtaining statistically meaningful estimates of concentrations of specific organisms per unit volume in compliance testing, as required by the IMO or proposed by the USCG. Given these limitations, the Panel recommends adding to future compliance protocols parameters that are much more rapidly and easily assessed. Examples of candidate "surrogate parameters" are shown in Table 6-4. They can be calibrated with organism numbers in laboratory tests on microcosm "ecosystems," but would be much more difficult, if not impractical, to calibrate for use with unknown types and numbers of organisms in ballast tanks. Therefore, surrogate parameters could be useful as bulk measurements in compliance testing. These parameters also could be used to augment existing measurements in land-based and shipboard verification testing. It will be critical to carefully calibrate all potential surrogate parameters with natural populations of ballast water flora and fauna before they can be used to evaluate the performance of BWMS – especially at the resolution of very low organism densities.

Increased Use of Tests at Multiple Spatial Scales

Instead of relying solely on full ship-scale testing, for practical reasons the Panel recommends that testing be conducted at a combination of scales as needed to address particular issues. Such tests would be done by the vendor of the BWMS prior to validation testing. For example, full-scale tests can pose extreme practical and logistical limitations and/or high risk in efforts to assess the effectiveness of treatment systems in removing maximal densities of harmful organisms, or mixes of representative organisms within certain density ranges. These risks support the use of sized-down mesocosm treatments (hundreds to thousands of liters; Ruiz et al. 2006; MERC 2009a-c) that are larger and therefore more realistic than bench-scale microcosms, but more manageable in volume than ballast tanks.

Sized-down treatments help to reduce risks to human health safety and receiving aquatic ecosystems for testing treatment system effectiveness at removing toxic substances and residues that are part of the treatment process. As Ruiz et al. (2006, p.10) stated:

> Economy of small scale and ease of manipulating environmental variables and community assemblage at the laboratory and intermediate scales make it possible and practical to estimate if a ballast water treatment process and system is likely to be effective over the full range of physical [chemical,] and biological conditions expected in the field;...the same regime on a ship would prove logistically and financially very unwieldy. Thus, smaller scale tests demonstrate the treatment's performance and capacity across a wide range of relevant state variables....

This approach also allows more precise, controlled sampling during test trials (MERC 2009b). At larger scales, practical limitations restrict the number of conditions that can reasonably be tested, and testing is directed more toward ensuring functionality of the engineered system rather than understanding the treatment process under various conditions.

Small-scale (benchtop or laboratory) experiments minimize logistics and expense, and they can provide proof of concept in assessing whether a given treatment meets expectations (Ruiz et al. 2006). For example, if a BWMS is planned for use across the salinity gradient, then its efficacy should be tested across all three salinity ranges (Table 6-2). Logistically, however, it may be feasible to test two salinity ranges at full scale, but not the third. In such cases, small-scale and intermediate-scale (see below) tests could be completed using the third salinity range. Likewise, the Panel recommends that bench-top and mesocosm experiments complement full-scale testing.

Table 6-4. Examples of candidate "surrogate" parameters for quantifying viable organisms in ballast water, and an analysis of their utility considering methods that are presently available*

Parameter(s)	Description	Suitability, Considering Methods Presently Available
Chlorophyll *a* (chl*a*)	"Universal" plant pigment, found in all phototrophic algae[1]; widely used as an indicator of total algal biomass[2,3]	*Pros*: Allows rapid processing of large numbers of samples; standardized methods widely available[2-5]. *Cons*: Cannot discern cell numbers per unit sample volume; not sensitive enough to detect < 10 cells ml-1 of small algae[6]; cellular chl*a* content highly variable (0.1-9.7% fresh weight) depending on the species[7] and the light conditions[8,9]; methods, and results depending on the method, vary widely[10-12]. *Present status*: Available methods do not allow reliable calibration with algal cell numbers in natural samples; improved methods are needed.
"Signature" or marker pigments	Diagnostic for cyanobacteria (zeaxanthin) and major eukaryotic algal groups (e.g. diatoms, other heterokontophytes - fucoxanthin; dinoflagellates - peridinin; chloro- phytes and euglenophytes - chl *b*)[3,13]	*Pros*: Potentially superior to chl*a* as algal biomass indicators; more specificity to algal groups (divisions or classes); standardized methods available[13-16]. *Cons*: Techniques must be applied carefully to avoid artifacts and sample bias[17]; low taxonomic resolution (can sometimes be improved by screening samples using microscopy[18] to identify abundant taxa5). *Present status*: Available methods do not allow reliable calibration with algal cell numbers in natural samples; improved methods are needed
Adenylates, especially ATP (adenosine triphosphate), and Total adenylates (ATP + ADP, adenosine di-phosphate, + AMP, adenosine mono- phosphate)	Indicator of total microbial biomass in plankton, sediments[17]	*Pros*: ~All microbial taxa have a ~constant ratio of ATP to total cell carbon[18]; easily extracted from microbial assemblages; not associated with dead cells or detritus[19,20] *Cons*: Cell ATP content varies for cells under environ- mental stress[21,22]; encysted cells with low metabolic activity have low ATP content (difficult to detect); total adenylates considered a better indicator of microbial biomass than ATP[17] within a given size class, but extrapolation from small sample volumes would lead to large error factors in estimating organism numbers. *Present status*: Available methods do not enable accurate assessment of small numbers of viable organisms per unit volume in stressed conditions within ballast tanks; improved methods are needed.

Parameter(s)	Description	Suitability, Considering Methods Presently Available
INT (2-*p*-iodophenyl)- 3-(*p*-nitrophenyl)-5-phenyltetrazolium chloride	Commonly used tetrazolium salt used to measure microbial activity (electron transport chain activity indicating viable organisms) in surface waters, biofilms, and sediments (freshwater to marine)[23-25]	*Pros*: INT accepts electrons from dehydrogenase enzymes and is reduced to a reddish colored formazan (INTF) - can be quantified by simple colorimetric analysis after a very short incubation time[26]; total cell numbers are quantified under epifluorescence microscopy using a counter-stain (e.g. acridine orange[27]) - proportion of population that is metabolically active is then estimated; very sensitive method even at low microbial biomass and low temperatures[27], with resolution at the level of individual cells28. *Cons*: Can miss cells with very low respiration (e.g. cysts), or cells that do not use INT as an electron acceptor[28]; still requires microscopy (tedious, time-consuming). *Present status*: Shows promise for use with various size classes of microorganisms in ballast water.
RNA, DNA	Quantitative PCR and related techniques; molecular and genomic probes[29-31]	*Pros*: Reliable quantification of targeted taxa from environmental water samples if PCR inhibitors can be removed and molecular material can be efficiently recovered[29,30]. *Cons*: More research is needed to test the degree to which these can reliably discern between viable and non-viable cells, or infective and non-infective cells, or toxic and nontoxic cells, unless supplemented by other techniques[31-34]. *Present Status*: Available methods do not allow reliable calibration with living algal cell numbers in natural samples; quantitative methods are emerging[35,36].

* References used: [1] Graham et al. (2009); [2,3,4,5] Wetzel (2001), Jeffrey et al. (1997), US EPA (1997), Sarmento and Descy (2008); [6] MERC (2009c); [7] Boyer et al. (2009); [8,9] U.S. EPA (2003), Buchanan et al. (2005); [10-12] Bowles et al. (1985), Hendrey et al. (1987), Porra (1991); [13] Schlüter et al. (2006); [14,15,16] Mackey et al. (1996), Jeffrey and Vest (1997), Schmid et al. (1998), Schlüter et al. (2006); [17] Sandrin et al. (2009); [18] Karl (1980); [19,20] Holm-Hansen (1973), Takano (1983); [21,22] Inubushi et al. (1989), Rosaker and Kleft (1990); [23-25] Songster-Alpin and Klotz (1995), Posch et al. (1997), Blenkinsopp and Lock (1998); [26] Mosher et al. (2003); [27]Sandrin et al. (2009); [28] Posch et al. (1997); [29] Caron et al. (2004); [30]Kudela et al. (2010); [31]Karlson et al. (2010); [32]Guy et al. (2003), [33]Audemard et al. (2004), [34]Burkholder et al. (2005), [35]Jones et al (2008), [36]Bott et al. (2010).

Practical Limitations of Challenge Water Conditions

While it is important to evaluate BWMS under diverse and challenging biological, physical and chemical conditions to understand system performance and the broad applicability and reliability of BWMS, all biological and chemical challenge conditions may not be achievable during a series of tests using natural waters. As described above, artificial manipulations of biological, physical and chemical conditions may introduce significant artifacts. Therefore, without rigorous validation that a modification to challenge water is representative of natural conditions and does not cause artifacts (e.g., stressing or killing organisms), only natural ambient conditions should be used. The following options are available to address the difficulty in meeting challenge conditions using un-augmented challenge water:

- Make the challenge conditions somewhat less stringent by allowing challenge water conditions in some replicate trials during testing of a given BWMS to fall outside of target values. For example, accept test results if all challenge water conditions for some replicate tests are within 70 percent of target values, as long as more than half the replicate trials are above all threshold values;
- Loosen the requirement that all the biological, physical and chemical challenge water conditions must be met for each replicate test of a given BWMS, as long as the majority of threshold conditions are met for an individual replicate test; or
- Allow certain manipulations or alterations to challenge water during testing of BWMS if approved by the EPA or the Verification Organization. Acceptance of the conditions would be based on either the test facility's test data or experimental data from other test facilities, showing the manipulations do not affect the validity of the test.

6.2.4. Testing Shipboard Treatment Systems: Inherent Mismatch between Viability Standard and Practical Protocols

The previous section reviewed features of current procedures for testing BWMS that could be improved with existing knowledge and technology. In this section, the Panel reviews additional aspects of current procedures that may not accomplish the stated goals because of inherent limitations in current knowledge and technology. All of the six issues considered below stem from the difficulty—perhaps the impossibility, given current technology—of accurately enumerating only those organisms that are viable (living). Current practices result from trying to directly assess the legal standards (which focus on viable organisms). This section is aimed especially at organisms < 50 μm, because the challenge of determining viability of larger organisms may be secondary to the problem of sampling an adequate volume to assess the concentration aspect of the legal standard (see section 3, Statistics and Interpretation). The Panel recommends that new approaches be developed, including procedures that address the standards indirectly, but have the benefit of practicality. In general, the Panel recommends that the limitations of testing protocols for determining "viability" and/or "living" be assessed. Where they are found to be lacking, the Panel recommends development of improved standardized protocols. If indirect metrics can be reliably correlated with the concentration of viable organisms at land-based test facilities, they should also be considered for adoption (as the Panel noted in Table 6-4 above).

As Lee et al. (2010, p. 72) aptly state, "A discharge standard of 'zero detectable organisms' may appear very protective; however, the true degree of protection depends on the sampling protocol." Here, a viable or living organism is defined as in U.S. EPA (1999), namely, as an organism that has the ability to pass genetic material on to the next generation. The percentage of non-viable cells can vary markedly, for example, from 5-60 percent among phytoplankton taxa, and in general, non-viable organisms are believed to represent a substantial component of the total plankton (Agusti and Sánchez 2002). There are several fundamental problems with present attempts to quantify viable organisms to evaluate ballast water treatment efficiency, outlined as follows.

Death of Organisms by Rapid Concentration from Large Volumes

A major issue confounding the realistic representation of viable organism concentrations is that the rapid concentration of organisms from large volumes into small volumes (which is a necessary prerequisite of enumeration) causes the death of many organisms across size classes. This concentration step must be accomplished quickly before organisms die, e.g., within six hours (or less) for zooplankton. Because of the need to evaluate relatively large volumes of water in order to be confident about the concentrations of sparse organisms in treated water, there is a fundamental disconnect in these requirements. It is difficult if not impossible to rapidly concentrate microflora and microfauna from relatively large volumes (hundreds of liters for zooplankton; liters for protists) by available filtration or centrifugation techniques without killing some of the organisms (e.g., Turner 1978; Cangelosi et al. 2007). Such rapid concentration techniques can cause the loss of a major fraction of the viable organisms, even when dealing with small sample volumes such as 1 liter (Darzynkiewicz et al. 1994). This problem affects zooplankton and protist size classes, especially delicate species such as wall-less algal flagellates. Thus, even if viable organisms can be distinguished from dead organisms when counted, what cannot be known is the proportion of the dead organisms that were actually living at the time of sampling. These problems illustrate the critical importance of a test facility validating all steps of any method it uses. It should be noted that concentration-related losses do not affect the smallest size classes, bacteria and viruses, because they are so abundant in most fresh, estuarine, and marine waters that it usually is not necessary to concentrate them from whole water samples prior to analysis by standard microbial techniques (U.S. EPA 2010; see below).

Organism Viability is Difficult to Determine

Organism viability is not easily detected by a single morphological, physiological, or genetic parameter, making it advantageous to use more than one approach (Brussaard et al. 2001). In the context of meeting a numeric standard, this approach is problematic because more than one 'answer' is generated. Furthermore, the procedures used to determine viability are specific to some taxonomic groups (e.g., vital stains) and have varying degrees of uncertainty in categorizing live versus dead. Even procedures recommended in the Protocol for land-based verification testing have practical limitations because of time constraints. For example, the Protocol defines dead zooplankton operationally as individuals that do not visibly move during an observation time of at least 10 seconds. Since live zooplankton may not move over that short period, death is verified by gently touching the organism with the point of a fine dissecting needle to elicit movement. However, the Protocol acknowledges that if every apparently dead zooplankter in a concentrated subsample was probed and monitored for at least 10 seconds, the length of time to complete analysis of the sample could be extended enough to increase the potential for sample bias due to death of some proportion of individuals that had survived the sampling and concentration procedures. In other words, the currently applied methods have serious limitations in some situations. The results are therefore most appropriately viewed as an index of the number of viable organisms.

The Protocol is a living document, and as better methods are developed, they will be incorporated. The Panel recommends consideration of a wider variety of indices that have the potential to be more rapidly completed, if not more accurate. These include parameters that may be correlated to the abundance of viable organisms, as discussed in the previous section (Table 6-4), and techniques to distinguish living from dead individuals prior to enumeration by other methods (e.g., microscopy). The latter is elaborated upon here.

Fluorescent stains have shown promise in detecting some live organisms or groups. For example, the fluorogenic substrate Calcein-AM (Molecular Probes Inc.) is used to stain live cells that have metabolic esterase activity (Kaneshiro et al. 1993; Porter et al. 1995). Once the colorless, nonfluorescent substrate is inside the living cell, its lipophilic blocking groups are cleaved by nonspecific esterases to a charged green fluorescen product that cannot pass across the plasma membrane. Dead cells cannot hydrolyze the Calcein-AM or retain the fluorescent product. Use of FDA, sometimes in combination with CMFDA (Table 6-3), is based on measuring intracellular esterases in live cells (Laabir and Gentien 1999; Hampel et al. 2001). FDA was described as a reliable, efficient method to quantify concentrated viable freshwater organisms in the > 10 to < 50 μm size class from ballast discharge (Reavie et al. 2010). However, various algal species differ in their uptake of FDA and CMFDA, and other particles in a given sample can also fluoresce (Garvey et al. 2007; MERC 2009c). The vital stain propidium iodide (PI), in combination with molecular probes, has been used to discern live from dead bacteria (Williams et al. 1998), but the number of false positives can vary widely (Steinberg et al. 2010), and this stain cannot be used to assess algal viability because its emission spectrum overlaps that of chlorophyll (Veldhuis et al. 2001). Consistent with the ETV Protocol (U.S. EPA 2010), the Panel recommends completion of on-site validation before selecting a viability method, including evaluation of false positives and false negatives.

As mentioned above, detection of infective viruses has received relatively little attention in ballast water treatment. Waterborne illnesses can involve a wide array of viruses; for example, enteric viruses that can be transmitted by water include poliovirus, coxsackievirus, echovirus, human caliciviruses such as noroviruses and sapoviruses, rotaviruses, hepatitis A virus, and adenoviruses (Howard et al. 2006). Considering pathogens of aquatic organisms, aquatic ecosystems are poorly understood with respect to the diversity of viral pathogens of beneficial aquatic life (Suttle et al. 1991; Griffin et al. 2003; Munn 2006; Suttle 2007). Viruses also cross size classes; in environmental samples, many are in range of nanometers, but some can be nearly 3 μm in maximum dimension (Bratbak et al. 1992); their tendency to adsorb to sediment particles (> 6 μm) means they can be captured with larger particles during sample preparation involving filtration (Bosch et al. 2005).

The U.S. EPA (2001) requires a 99.9 percent reduction in the total number of human enteric viruses in water for human consumption. In practice, this requirement is met based on treatment alone, although the EPA acknowledges that removal actually can only be accurately assessed by monitoring finished waters over time. Ultra-filtration protocols have been developed for concentrating and enumerating human enteric viruses (Fout et al. 1996; U.S. EPA 2001), but these techniques do not discern potentially infectious from non-infectious viruses. Environmental water samples also have been evaluated for human viral pathogens using standard techniques for *in vitro* cultivation, an approach that is affected by the same problems confronted for detection of viable bacteria—the techniques are expensive, time-consuming, labor-intensive, and can easily miss various groups of infectious viruses (Fout et al. 1996). Rapid, sensitive molecular methods for viral nucleic acid detection have been recently developed but, again, most cannot discern potential infectivity. The intercalating dye propidium monoazide (PMA) has shown promise in detecting potentially infective coxsackievirus, poliovirus, echovirus, and Norwalk virus (Parshionikar et al. 2010). In other promising research, Cromeans et al. (2005) included additional processing steps such as specific capture by cell receptors for Coxsackie B viruses *in vitro*, followed by molecular detection of viral nucleic acids in the captured viruses; or selection/detection of specific RNA present in host cells only during virus replication. Real-time assays (30-90 minutes) were also developed for enterovirus, hepatitis A virus, Adenovirus, and Norovirus detection. There remains a need for new, commercially available technology that can discern infectious from non-infectious viruses (Cromeans et al. 2005; Parshionikar et al. 2010) although current and proposed standards do not distinguish the two.

Special Challenges of Resistant or Nonculturable Stages in Attempts to Assess Viability

Resting stages (e.g., cysts) of some bacteria, phytoplankton, protists, zooplankton and metazoans are particularly resistant to motility, staining, and any other tests. For example, the protist size class ($\geq$ 10 μm to < 50 μm) includes many species (microalgae, heterotrophic protists, metazoans) that form dormant cells or resting stages, or cysts (Matsuoka and Fukuyo 2000; Marrett and Zonneveld 2003). Cysts from potentially toxic dinoflagellates are commonly found in ballast waters and sediments (Hallegraeff and Bolch 1992; Dobbs and Rogerson 2005; Doblin and Dobbs 2006). These cysts have been used as model indicator organisms to assess ballast water treatment efficiency (Anderson et al. 2004; Stevens et al. 2004), based on the premise that treatments which can eliminate the cysts likely also eliminate other, less-resistant organisms (Bolch and Hallegraeff 1993; Hallegraeff et al. 1997).

Because resistant cells often have a low metabolic state and thick, multi-layered walls that are impermeable to many stains (Romano et al. 1996; Kokinos et al. 1998; Connelly et al. 2007), their viability can be difficult to assess without culture analyses that may require weeks to months (Montresor et al. 2003; U.S. EPA 2010). Improved methods have been developed for some algal groups (Binet and Stauber 2006; Gregg and Hallegraeff 2007) but, overall, as the ETV Protocol (U.S. EPA 2010, pp.46-47) states, “At present, no rapid, reliable method to determine cysts’ viability is in widespread use, and the FDA-CMFDA method has yielded variable results with dinoflagellates and cyst-like objects.” The ETV Protocol recommends use of this method as a “place holder” until more effective methods become available.

The effectiveness of ballast water treatment in removing viable bacteria is evaluated by using multiple bacterial media in combination with taxon-specific molecular techniques (MERC 2009c; U.S. EPA 2010 and references therein). Colonies are monitored and quantified after ~1 to 5 days, depending upon the organism and its growth. These methods enable detection and quantification of viable, culturable cells. However, it has been repeatedly demonstrated that bacterial consortia across aquatic ecosystems commonly have a substantial proportion of cells that are active (viable) but nonculturable (Oliver 1993; Barcina et al. 1997 and references therein). These cells obviously would be overlooked in culturing techniques, a problem that would result in failure to detect viable cells of bacterial pathogens in treated ballast water. Under some conditions, the nonculturable organisms can regain activity and virulence (Barcina et al. 1997 and references therein).

Biased Counts Due to Live, Motile Species Changing their Location in Counting Chambers

At the other extreme from resting stages are living organisms that are difficult to enumerate because they are highly mobile. Organisms are typically enumerated in counting chambers, based upon an underlying premise that the cells do not change their location in the chamber. However, many protists move rapidly by means of flagella or other structures. Because they do not maintain their position in a counting chamber, as live cells they could be counted multiple times. Moreover, their sudden movement can disrupt the locations of other cells in the chamber, mixing cells that may have been counted with others that have not yet been counted. For these reasons, reliance on live counts can easily yield unreliable data. This consideration underscores the need for vigorous validation of protocols used to quantify viable organisms.

Indirect Metrics for Enumeration of Viable Cells Should be Investigated for Use in Standard Protocols

Consideration of the above points—death during concentration of organisms, lack of reliable procedures to assess viability (especially for resting stages of many taxa), movement of live organisms in counting chambers that can result in serious quantification errors—leads the Panel to recommend that alternative approaches, including enumeration of preserved organisms and indirect metrics of the concentration of viable organisms, be tested. Should they be validated as superior to present protocols, then the argument is strong to consider elevating these alternative approaches to standard protocols. These inherent limitations add weight to the more practical considerations in section 6.2.3 above: the practical and inherent limitations converge as an argument for the greater development, testing and implementation of indirect metrics of the concentration of viable organisms, including both STOs and surrogate parameters, particularly in compliance testing. Adding parallel testing of indirect metrics to land-based testing currently underway in test facilities from different geographic regions could rapidly yield comparisons on which decisions for future testing could be made. Possibly, a combination of approaches will prove to be the most advantageous in estimating the concentration of viable organisms of different taxonomic groups.

6.3. Approaches to Compliance/Enforcement of Ballast Water Regulations and Potential Application to Technology Testing

The EPA has extensive experience in effective compliance and enforcement of discharge regulations, and has committed to work with the USCG to develop and implement compliance and enforcement measures for ballast water regulations (2011 MOU between EPA and USCG[5]). However, given the nature of ship ballast water discharge, new approaches likely will be needed. Both initial testing of treatment systems (6.2.2 – 6.2.4) and methods currently available for potential compliance and enforcement monitoring are complex, slow and expensive. Statistical (see section 3, this report) and logistical limitations related to collection of appropriate sample volumes and detection/quantification of live organisms in practice, mean that it may often be impossible to directly assess whether a vessel can meet all the numerical standards for viable organisms (King and Tamburri 2010). No information was provided to the Panel on whether protocols and systems for compliance monitoring (whether voluntary by ship operator or legally required) and enforcement were being considered alongside the development and testing of treatment systems. The Panel considers it essential that these be developed so enforcement can commence as soon as a U.S. ballast water performance standard is finalized.

The practical and inherent limitations suffered by the full protocols for verification testing of BWMS (6.2.2 – 6.2.4) have even greater force in the context of routine inspections (either self-inspections or regulatory inspections) (King and Tamburri 2010). They are simply not practicable to use in the compliance and enforcement context. Unless alternative protocols that are practical for inspections are developed, neither self-compliance efforts nor regulatory enforcement will be possible once a system is installed on a ship. For example, treatment system malfunctions are inevitable. If some types of mechanical failure are not obvious to the operator or inspector, release of organisms may reach and maintain non-compliant levels for long periods of time with no detection of the malfunction, no penalty, and therefore no incentive to detect and fix the system. Unenforceable rules are bound to fail to meet the goal of reducing invasions. Therefore, the Panel recommends that the EPA develop an approach for BWMS that includes metrics appropriate for compliance monitoring and enforcement.

A potential solution is the use of a step-wise compliance reporting, inspections, and monitoring approach, described below, which involves a series of steps that increase the likelihood of detecting non- compliance but also increase in cost and logistic challenges (King and Tamburri 2010).

- *Reporting.* Vessel owner or ship master submits reports on the type of certified treatment system onboard and documentation demonstrating appropriate use and maintenance.
- *Inspections.* Enforcement official boards vessel and inspects the certified treatment systems to verify use and appropriate operations and maintenance.
- *Measures of system performance.* Indirect or indicative water quality measures are collected autonomously (using commercially available instruments), or by inspectors, that demonstrate appropriate treatment conditions have been met.
- *Indirect measures of non-compliance.* Indirect metrics of abundances of live organisms are collected autonomously, or by inspectors, for indications of clear non-compliance.
- *Measures of performance standard.* Direct measures of concentration of live organisms in the various regulated categories are made by specially trained technicians, with statistically appropriate sampling and validated analyses and methodologies.

Protocols assessing indirect surrogate measures to quantify viable organisms should be further developed for quick, easy, and defensible shipboard compliance monitoring (see 6.2.2 - 6.2.4).

6.4. Reception Facilities as an Alternative to Shipboard Treatment

Proposed federal regulations and the IMO Ballast Water Management Convention allow for the transfer of vessel ballast water to reception facilities, where the organisms in ballast water would be removed or inactivated. Various studies have envisioned reception facilities as either built on land or installed on port-based barges or ships. The discussion here refers to the use of on-land facilities, unless otherwise specified.

Ballast water would be pumped off a vessel to a reception facility through main deck fittings and piping or hoses similar to those currently used to transfer oil or other liquid cargo, oil-contaminated ballast water, or fuel oil between vessels and shore. Vessels would need to be outfitted with appropriate pipes and pumps to move ballast water to the deck and off the ship at a fast enough rate so the vessel is not unduly delayed in port. The reception facility would store and treat the ballast water before discharging it to local waters.

Vessel architecture and operations are principal impediments to the development of shipboard BWMS. Challenging factors include vibration, small and busy crews, limited space and weight allowances, limited power, potentially increased corrosion rates and sometimes short voyages. Reception facilities, relieved of many or all of those constraints, show promise to achieve more stringent ballast water treatment standards than shipboard BWMS.

The Panel did not reach consensus on certain issues and analyses related to treatment of ballast water at reception facilities. Issues on which the Panel reached consensus are described in sections 6.4.1 and 6.4.2 (although two points of view are included for one of the issues). Panel conclusions on reception facilities are presented in section 6.7.

6.4.1. Potential of Reception Facilities to Cost Effectively Meet Higher Standards

Though various studies, regulations and guidelines recognize the potential of reception facilities to treat ballast discharges, the EPA and USCG reports on ballast water treatment have not addressed reception facilities (EPA 2001; Albert et al. 2010; USCG 2008a,b). The literature on onshore treatment is reviewed in Appendix B. Some studies conclude that reception facilities are a technically feasible option for either the industry as a whole or for some part of the industry (Pollutech 1992; NRC 1996; Oemke 1999; CAPA 2000; California SWRCB 2002; Brown & Caldwell 2007, 2008). Other studies conclude that cost or other factors could limit the use of reception facilities to part of the industry (Victoria ENRC 1997; Dames & Moore 1998, 1999; Rigby and Taylor 2001a, b; California SLC 2009, 2010).

Four studies compared the effectiveness or costs of reception facilities and shipboard treatment. Pollutech (1992) ranked reception facilities second in terms of effectiveness, feasibility, maintenance and operations, environmental acceptability, cost, safety and monitoring out of 24 ballast water management approaches for Great Lakes vessels; this study ranked shipboard filtration through a 50 μm wedgewire strainer higher, and 17 other shipboard treatments lower, than treatment by a reception facility[6]. AQIS (1993a) found reception facilities (considering both on land and barge-mounted facilities) to be less expensive than shipboard treatment in both single-port and nation-wide scenarios in Australia, concluding that reception facilities "are more economic and effective than numerous shipboard plants." Aquatic Sciences (1996) estimated the costs of using barge-mounted reception facilities in the Great Lakes, and concluded that it is technically feasible, "more practical and enforceable" than shipboard treatment, and offers "the best assurance of prevention of unwanted introductions." California SWRCB (2002) found reception facilities to be the only approach to have acceptable performance in all three categories of effectiveness, safety, and environmental acceptability in a qualitative comparison with 10 shipboard treatments. Cost estimates compiled by the U.S. Coast Guard (e.g., USCG 2002) showed reception facilities to be generally less expensive on a per metric ton basis than shipboard treatment, although these estimates predate the establishment of discharge regulations and the most recent generation of BWMS. In fact, most of the existing studies and estimates use outdated assumptions or data, or are based on specific regions; therefore their conclusions may not apply to the current U.S. situation, nor do they address international shipping issues.

The potential advantages of reception facilities over shipboard treatment systems include: fewer reception facilities than shipboard systems would be needed; smaller total treatment capacity would be needed; and reception facilities would be subject to fewer physical restrictions, and would therefore be able to use more effective technologies and processes such as those commonly used in water treatment. A shift from shipboard treatment to reception facilities is in some ways analogous to a shift from household septic tanks to centralized wastewater treatment plants. These advantages are discussed in greater detail in the following paragraphs.

Treatment Capacity

The EPA estimates that approximately 40,000 cargo vessels and 29,000 other vessels will be subject to ballast water discharge requirements in the U.S. over the five-year VGP period (Albert and Everett 2010); approximately 7,000 ocean-going vessels called at U.S. ports in 2009 (MARAD 2011). Using reception facilities would reduce the number of treatment plants and the total treatment capacity needed for ballast water management. In shipboard treatment, a plant is installed on each vessel, and for nearly all types of BWMS this must be large enough to treat the vessel's maximum ballast uptake or discharge rate (Lloyd's Register 2010). The total treatment capacity needed is thus equal to the sum of the maximum uptake or discharge rate of all ships. In contrast, reception facilities serve a number of vessels, and since all vessels do not arrive and discharge ballast water simultaneously, the treatment capacity needed would be less. Ballast water storage tanks at reception facilities would further lower the needed treatment capacity, potentially to the average ballast water discharge rate. However, existing studies do not appear sufficient to reliably estimate total treatment capacities required for individual BWMS versus a national U.S. network of reception facilities. If undertaken, such studies would benefit from explicit statements of assumptions about what drives treatment capacity needs and from comparisons of capacity estimates derived from a range of assumptions.

Constraints on Treatment

Constraints on onboard treatment include limited space, power and treatment time, and ship stability challenges (Pollutech 1992; AQIS 1993a; Aquatic Sciences 1996; NRC 1996; Cohen 1998; Oemke 1999; Reeves 1999; California SLC 2010; Albert and Everett 2010). These constraints are largely absent in reception facilities,

Efficacy of Treatment Methods

Any treatment used on vessels could be used in reception facilities; alternatively, there are methods available for reception facilities that cannot be used on vessels because of the space and other constraints listed above. Such technologies include common water or wastewater treatment processes, such as settling tanks and granular filtration, and less common processes including membrane filtration (AQIS 1993a; Gauthier and Steel 1996; NRC 1996; Victoria ENRC 1997; Reeves 1999; Cohen and Foster 2000; California SWRCB 2002; California SLC 2010).

The following information illustrates what could be achieved by using available water/wastewater technologies in reception facilities; it describes what can be achieved in drinking water treatment systems, recognizing that reception facilities would have to deal with a much greater taxonomic diversity of organisms (from large zooplankton to microbes) and be able to effectively treat all possible salinities (not just freshwater). For example, testing protocols for shipboard BWMS require at least a four-log reduction in the ≥ 50 um size class, and at least a two-log reduction in the ≥ 10 to < 50 µm size class when compared to ballast water uptake conditions. These metrics, however, do not account for organism mortality in ballast water that occurs even with untreated ballast water. As this mortality varies significantly (in some cases resulting in one-log reductions), it is difficult to quantify efficacy in terms of log reductions. As such, Table 6-6 below compares the level of treatment that would be required for two discharge standards relative to mean organism counts taken from vessels after a voyage. The Panel recognizes that there may be valid ways of assessing efficacy other than basing it on mean concentrations.

EPA requires that drinking water treatment systems be capable of at least 3-5 log reductions in *Giardia*[7], 3-5.5 log reductions in *Cryptosporidium*[8], and 4-6 log reductions in viruses, depending on the source water (US EPA 1991, 2006). Several common drinking water filtration technologies are capable of 3-4 log reductions in protozoans and bacteria and 2-4 log reductions in viruses, and membrane filtration can achieve >4-7 log reductions (U.S. EPA 1991, 1997b; NESC 2000a; LeChevallier and Au 2004; Wang et al. 2006; WHO 2008). UV disinfection can achieve 2-3 log reductions in protozoans and 3-4 log reductions in bacteria and viruses; biocides can achieve at least three-log reductions in *Giardia*, 3-6 log reductions in bacteria, and 3-4 log reductions in viruses depending on dose and contact time (U.S. EPA 1997b; Sugita et al. 1992; NESC 2000b; LeChevallier and Au 2004). Filtration and disinfection are generally considered additive processes: that is, a filtration process can produce a 3 log reduction, and a disinfection process can produce a two-log reduction, and in sequential combination could potentially produce a five-log reduction (U.S. EPA 1991).

Thus, even without a disinfection step, it appears that several common drinking water filtration technologies available for reception facility use could achieve the 1.5-2.7 log reductions from mean ballast water concentrations needed to meet the IMO D-2 and USCG Phase 1 standards, although this has not been tested with ballast water. It has not been demonstrated that these technologies could address the extremely high numbers of organism found in the ballast water of vessels after some voyages. Several combinations of filtration plus a single disinfection process appear to be able to achieve the 4.5-4.9 log reductions needed to meet the USCG Phase 2 requirements for viruses, bacteria and organisms in the ≥ 10 to < 50 µm size class, and perhaps also the 5.7 log reduction needed to meet the USCG

Phase 2 standard for organisms ≥ 50 μm Treating with one or more additional disinfection processes could produce greater log reductions.[9]

Some membrane filtration technologies that could be used in reception facilities have produced results of no detectable organisms in different organism classes. For example, the microfiltration unit used in the conceptual design for a reception facility at the Port of Milwaukee (Brown & Caldwell 2008) would likely result in no detectable organisms in both the ≥ 50 μm and ≥ 10 to < 50 μm size classes (based on microfiltration results cited in U.S. EPA 1997b and LeChevallier and Au 2004). On the other hand, ultrafiltration or nanofiltration might be needed to leave no detectable bacteria or viruses in the effluent, although the time required to filter water to this level and its effect on vessel operations has not been evaluated.

Plant Operation by Trained Water/Wastewater Treatment Personnel

Shipboard BWMS likely would be operated and maintained by regular crew members as added duties (NRC 1996; California SLC 2010). Studies have noted that many of these crews are already overburdened. Operation by trained, dedicated personnel in reception facilities would likely result in more reliable performance (Cohen 1998; California SWRCB 2002; Brown & Caldwell 2007; California SLC 2010). Maintenance and repair work are more likely to be done reliably, and replacement parts obtained more quickly, in reception facilities (AQIS 1993a; Aquatic Sciences 1996; Cohen 1998).

Safety

Restricted working spaces and difficult or hazardous working conditions at sea (AQIS 1993a; Cohen 1998; Cohen and Foster 2000) increase the risk of accidents with shipboard treatment. The storage and use of biocides or other hazardous chemicals pose greater risks to personnel on vessels than in reception facilities (AQIS 1993a; Carlton et al. 1995; Reeves 1998; Cohen 1998) and greater risk of accidental discharge to the environment (Pollutech 1992; AQIS 1993a; Carlton et al. 1995). Because some physical treatment processes cannot be used onboard, shipboard systems might rely on biocides more than would reception facilities.

Table 6-6. Log reductions required by different discharge standards

Discharge standard	Organism/size class:			
	≥ 50 μm	≥ 10 - < 50 μm	Bacteria	Viruses
	per m^3	per ml	per ml	per ml
IMO D-2 and USCG Phase 1	2.7	1.5	no reduction	no reduction
USCG Phase 2	5.7	4.5	4.9	4.9

Reductions are from mean values reported by IMO (2003) for unexchanged and untreated ballast water sampled from vessels at the ends of voyages, for zooplankton (n=429, collected with 55-80 μm mesh nets, corresponding approximately to organisms in the ≥50 μm size class), phytoplankton (n=273, collected with <10 μm mesh sieves or counted in unconcentrated samples, corresponding approximately to organisms in the 10-50 μm class), bacteria (n=11) and virus-like particles (n=7).

On the other hand, increased safety risk and a risk of spills or leaks of untreated ballast water may accompany transfers of ballast water to reception facilities. Although liquid transfer is common practice for tank ships, many other ships do not have crews experienced in these operations, and safety training would be needed.

Reliability

Operation and maintenance by dedicated wastewater treatment staff should make the reliability of reception facilities greater than that of shipboard BWMS. Extensive, long-term experience with water and wastewater treatment technologies provides a basis for estimating the expected long-term performance of these technologies if employed in reception facilities, while the brief and limited experience with shipboard BWMS provides little basis for assessing whether they are likely to perform adequately over a 20-to-30-year vessel lifetime. Since many BWMS treat ballast water on uptake (Lloyd's Register 2010), which many vessels hold in dedicated ballast tanks where cysts or other resting stages may be retained in sediments for long periods (Cohen 1998), failure to operate a BWMS or to operate it effectively at any time could contaminate treated ballast water on later voyages (AQIS 1993a; Reeves 1998). In addition, reception facilities would have more flexibility to build redundancy into the system design than would shipboard systems.

Adaptability

Because of space restrictions on vessels and structural cost factors that make treatment components a smaller part of the total cost of reception facilities, it is likely to be both physically and financially easier to retrofit, replace or upgrade reception facilities than shipboard systems. Reception facilities "provide treatment flexibility, allowing additional treatment processes to be added or modified as regulations and treatment targets change" (Brown & Caldwell 2008).

Compliance Monitoring and Regulation

Although the requirements for demonstrating compliance with ballast water discharge regulations have yet to be established, the effort and cost of monitoring and enforcement needed to meet a given standard could be much less for a small number of reception facilities compared to a larger number of mobile, transient, shipboard plants, most of which are foreign-owned or foreign-flagged, which are accessible only when in U.S. ports, usually for brief periods (AQIS 1993a; Ogilvie 1995; Aquatic Sciences 1996; Cohen 1998; Dames & Moore 1999; Oemke 1999; California SWRCB 2002; Brown & Caldwell 2007; California SLC 2010). Some studies noted that only reception facilities put the responsibility for monitoring, control and effectiveness entirely in the hands of the authorities responsible for protecting the receiving waters, without reliance on marine vessel logs or on authorities in originating ports (AQIS 1993b; Dames & Moore 1999; California SWRCB 2002).

6.4.2. Challenges to Widespread Adoption of Reception Facilities in the U.S.

Although reception facilities offer advantages as just discussed, the Panel recognizes that there are challenges to their adoption. The Panel reached consensus on all but one of the challenges presented in this section, and opposing views are presented for that one.

Ballast Discharge before Arrival to Reduce Time Spent at Berth

Some vessels may discharge part of their ballast water before arriving at berth so they can complete discharge by the time the cargo is loaded (AQIS 1993a; Oemke 1999; Cohen and Foster 2000; CAPA 2000; Rigby and Taylor 2001a). Alternatively, a ship's ballast water system can be outfitted with pipes and pumps that are large enough to allow the ship to unload ballast water as quickly as it loads cargo (AQIS 1993b). Glosten (2002) and Brown & Caldwell (2007, 2008) identified technical solutions for retrofitting a variety of vessels (but not all types) to allow them to deballast at berth during the time they load cargo. However, this issue has not been studied with respect to costs or feasibility for handling as yet uncertain estimates of the numbers of vessels expected to require treatment at different ports.

Ballast Discharge to Reduce Draft before Arriving at Berth

Several studies noted that some vessels discharge ballast water before arriving at berth to reduce draft to cross shallows (Cohen 1998; Dames & Moore 1998, 1999; Oemke 1999; CAPA 2000, Rigby and Taylor 2001a; California SWRCB 2002; California SLC 2010). The frequency of these occurrences has not been quantified (one authority the Panel consulted stated that they are rare whereas another indicated that some Great Lakes operators may perform such discharges routinely). Possible solutions include offloading ballast water to barges as is done for some liquid cargos (AQIS 1993a; Carlton et al. 1995; Dames & Moore 1999; CAPA 2000; Rigby and Taylor 2001a; Glosten 2002; California SWRCB 2002), or importing cargo in shallower-draft ships. Dames & Moore (1998) suggested that a barge- or ship- mounted reception facility could service deep-drafted arrivals that need to deballast during approach. Some panel members point out that this issue has not been studied with respect to costs or feasibility for handling as yet uncertain estimates of the numbers of vessels expected to require treatment at different ports.

Ballast Discharge by Lightering Vessels

Large tankers that arrive on the U.S. coast carrying crude oil or other liquid cargo may transfer part of it to lightering vessels (smaller tankers or barges) in designated anchorages or lightering zones. These lightering vessels often discharge ballast as they load cargo. In many cases, the discharged ballast water is from nearby sites (CDR Gary Croot, U.S. Coast Guard, pers. comm.; National Ballast Information Clearinghouse data), and depending on how the regulations are written may not require treatment.[10] In cases where the ballast water is from more distant sites, solutions might include offloading ballast water to lightering vessels that have been ballasted with local water, or importing cargo in smaller tankers. The frequency, volumes, and uptake and discharge locations associated with lightering have not been quantified, so the significance of this issue with respect to invasions or feasibility of technical solutions is not known.

Implementation Schedule

It typically takes up to 30 months to design, permit and construct a sewage treatment plant larger than 10 mgd, and potentially much longer if sites are scarce, or if there are issues related to permit approvals (Robert Bastian, U.S. EPA Office of Water, pers. comm.). Most ballast water reception facilities needed in the U.S. would be smaller. Vessel modifications are needed for either shipboard or reception facility approaches, either to install a BWMS or to allow rapid discharge to a reception facility. This process is almost exclusively undertaken while the vessel is out of service, which occurs infrequently; dry dockings, by marine vessel classification society requirement, must be no less than once every five years (ABS SVR 7/2/1-11). To accommodate vessel modifications, proposed standards include phase-in periods (8 years for IMO D-2, 9 years for USCG Phase 1). The critical path for both reception facility and shipboard treatment is the vessel modification work, where the governing factor is the frequency with which the vessel is taken out of service. This is the same for either approach.

A more comprehensive comparison of potential implementation schedules for both shipboard BWMS and reception facilities is needed.

The Current Regulatory Framework

Challenges associated with the regulatory framework are included in this section even though the Panel did not reach consensus on this issue, because many Panel members thought that this is the major challenge to reception facilities; therefore leaving it out would result in an unbalanced portrayal of advantages and challenges. Two views of the issue are presented here.

- *View 1*: Although reception facilities are allowed in policy and rules and have identified advantages relative to BWMS, there are no reception facilities currently available in the U.S. to remove organisms from ballast water. At the same time, there are 10 internationally Type Approved BWMS of which many have been sold. This appears to be a result of the framework of the 2004 IMO Convention that phases in performance standards by marine vessel ballast water capacity and construction date of marine vessels rather than on a port-by-port basis. To avoid the risk of arriving in a port without an operational reception facility, operators are opting to install shipboard BWMS. The U.S. proposed Phase 1 timetable would require all new vessels constructed starting in 2012 to meet performance standards upon delivery. To be in compliance using only reception facilities, the marine vessel operator must be assured that there will be an operational reception facility at all anticipated ports-of-call where ballast water discharge might be expected for the lifetime of the vessel. On the other hand, vessels engaged solely in regional trade may benefit from the reception facility approach if reception facilities are operational in the region and will not need to invest in a shipboard BWMS.
- *View 2*: The alternative view holds that current federal regulations governing ballast discharges under NISA and CWA are based on mid-ocean exchange (Albert et al. 2010), and thus favor neither BWMS nor reception facility treatment. Various states have adopted discharge standards, of which some might be met by BWMS, while others would require reception facility treatment because their requirements are more stringent. Regulatory agency dismissal of or opposition to reception facilities, and encouragement of BWMS (including the sponsorship and funding of research), has contributed to the focus on BWMS. Equipment manufacturers have invested in the development of BWMS because they expect that discharge standards that can be met by BWMS will be implemented and enforced, thereby creating a large enough market to allow them to recoup their investments and turn a profit. Ports have not promoted the development of reception facilities because they are not convinced that

discharge standards requiring treatment in reception facilities will be implemented and enforced effectively. If equipment manufacturers and ports come to believe that standards will be implemented that will need to be met by treatment in reception facilities—then the current focus on BWMS will shift. It is the decisions, actions and communications of regulatory agencies that will mold these expectations about the future direction and implementation of discharge standards.

Issues Regarding Treatment in Reception Facilities for Which the Panel Did not Reach Consensus

There were several additional issues regarding treatment in reception facilities for which the Panel did not reach consensus. Included below is a brief summary of discussions that may be helpful to the ballast- water community in the future.

- *Need for further study.* The Panel discussed the need for further study of BWMS treatment options. Some Panelists, noting the scarcity of reliable test data, discussed the need for further study of long-term performance of shipboard BWMS under the challenging conditions of actual shipboard use. Some Panelists discussed the need for pilot studies of reception facilities to assess their cost, operations, and safety issues in order to assess systemic challenges and to support operational solutions for creation of networks of onshore reception facilities.
- *Cost comparison.* The Panel did not reach agreement on issues relating to estimating and comparing the cost of treating ballast water in shipboard BWMS and the cost for treating in onshore reception facilities. There were differing opinions on the assumptions needed to develop screening estimates for either option. These included assumptions about capacity requirements, applicability of existing cost data, extrapolation methods, inclusion of operational costs that could be incurred by vessel owners if they were delayed due to unavailability of reception facilities or from inoperable BWMS, and costs for some vessel owners that might be required to install a shipboard BWMS as well as pay for use of a reception facility, depending on the port of call.
- *Implementation issues.* The Panel discussed issues that could affect the time needed to implement treatment of ballast water by individual shipboard BWMS or for developing a network of reception facilities, but did not come to agreement on their implications for implementation timelines. Some panel members said that reliance on shipboard BWMS would require a potential lag of several years for large-scale production of BWMS and time needed to develop effective monitoring and enforcement. Some members said that timelines for developing reception facilities would need to consider implications of land availability adjacent to port terminals, and time to acquire and permit newly designed treatment facilities and required support services.
- *International issues.* The Panel briefly discussed international issues related to potential new U.S. standards for ballast water discharge. These issues included the complexities of implementing standards for vessels engaged in international maritime trade. The Panel also briefly discussed whether setting U.S. standards based on the likely higher performance of reception facilities would introduce potentially

varying levels of protection against introduction of invasive species among the U. S. and other countries.

6.5. Approaches Other than Ballast Water Treatment

Several approaches other than the treatment of ballast water could help to reduce the risk of biological invasions from ballast water discharges, and contribute to the achievability of performance standards and permit requirements. While these approaches are often recommended, including by IMO, they are not often required or incentivized in practice. These approaches include ballasting practices to reduce the uptake of organisms, ballast water exchange to reduce the concentration of exotic organisms, reductions in the volume of ballast water discharged in U.S. waters, and management of the rate, pattern or location of ballast water discharge to reduce the risk of establishment. Although the charge questions to the Panel focused on shipboard treatment, the Panel considered these other approaches because, when used in combination with shipboard treatment, they appear to be capable of achieving a greater level of risk reduction than shipboard treatment alone.

6.5.1. Managing Ballast Uptake

Several studies have recommended various ballasting practices—sometimes referred to as ballast micro- management (Carlton et al. 1995; Oemke 1999; Dames & Moore 1998, 1999; Cohen and Foster 2000), shipboard management measures (Gauthier and Steel 1996), or precautionary management measures (Rigby and Taylor 2001a,b)—to reduce the number of organisms, or the number of harmful or potentially harmful organisms (such as bloom-forming algae and human pathogens found in sewage), that are taken up with ballast water It is suggested that this can be accomplished by managing the time, place and depth of ballasting. Some of these measures have been included in laws, regulations or guidelines, including IMO guidelines and the USCG rules implementing the National Invasive Species Act. Although some of these regulations or guidelines have been in effect for nearly 20 years, there appear to be no data on levels of compliance and no studies of the effectiveness of any of these measures in reducing the uptake of organisms.

While there may be reasons for skepticism regarding the effectiveness or feasibility of several of these measures (AQIS 1993b; Cohen 1998; Dames & Moore 1998, 1999; Cohen and Foster 2000; Rigby and Taylor 2001b), some could be helpful in meeting stringent standards if vessels had sufficient incentive to implement them. The effectiveness of alternative ballasting (e.g., at locations low in harmful organisms) and deballasting practices (e.g., locations and practices to reduce concentrating propagules) should be quantified. As an example of the former, researchhas shown that taking up ballast water in areas affected by toxic dinoflagellate blooms, followed by deballasting in another location, can result in distribution of those blooms to previously unaffected areas (Hallegraeff and Bolch 1991). Clearly, such action should be avoided as routine practice, and can also help to meet BWMS standards.

The value of such practices could be evaluated with models using currently available data on organism distributions or by experimental approaches. To the extent these practices would reduce the uptake of organisms, they could be used by vessels to help them meet any performance standards that might be adopted. From the perspective of overcoming technical limitations on the feasibility of meeting different performance standards, such practices might allow the adoption of -- and vessel compliance with -- more stringent standards than would otherwise be achievable. Thus, there are valid reasons for the EPA to consider the potential for employing these practices in combination with ballast water treatment to further reduce the risk of releasing exotic organisms in U.S. waters.

6.5.2. Mid-ocean Exchange

Mid-ocean ballast water exchange has the potential, in combination with the other approaches discussed here, to further reduce the concentration of exotic organisms (though not necessarily reduce the concentration of all organisms) in ballast discharges. There is general agreement that when properly done, ballast water exchange can reduce the concentration of initially loaded organisms by about an order of magnitude on average (Minton et al. 2005). It is not however, always possible, especially for short coastal voyages. Additionally, conducting exchange represents an additional cost to the vessel.

6.5.3. Reducing or Eliminating Ballast Water Discharge Volumes

Invasion risk is positively related to the total number of propagules released in a given time and place. Thus, risk is positively related to the concentration of propagules times the volume of the discharge. Even if the concentration of propagules is unmanaged, reducing discharge volumes will reduce invasion risk in ways that are predictable across taxa (Drake et al. 2005). Given this, various alternatives to the use of "conventional" ballast water management systems have been proposed and studied since the 2004 IMO Ballast Water Management Convention. These emerging alternatives to shipboard BWMS include concepts and designs for "ballastless" or "ballast-free" ships, "ballast-through" or "flow-through" ships, the use of "solid-ballast", and the use of "freshwater ballast". In fact, Regulation B-3, of the IMO Convention predicts and allows for the development and future use of such approaches to prevent the transport of invasive species by ships. These approaches are summarized below.

Ballastless ship designs constitute a fundamental paradigm shift in surface vessel design. Rather than increasing the weight of vessels by adding water to ballast tanks, these new designs use reduced buoyancy to get the ship down to safe operating drafts in the no-cargo condition. For example, the Variable Buoyancy Ship design (Parsons 1998; Kotinis et al. 2004; Parsons 2010) achieves this end by having structural trunks of sufficient volume that extend most of the length of the ship below the "ballast waterline" and then opening these trunks to the sea in the no-cargo condition. When the ship is at speed, the natural pressure difference between the bow and the stern induces flow through the open trunks, resulting in only local water (and associated organisms) within trunks at any point during a voyage. While showing promise, and worthy of further considerations, ballastless ship designs appear feasible only for new vessels being built in the future and may result in an overall increase in vessel biofouling (another significant source of invasive species), if surfaces in open flow-through spaces are more accessible and hospitable than traditional ballast tank surfaces (which are rarely fouled by higher organisms). Similarly, a return to a historic approach of using solid ballast (commonly iron, cement, gravel or sand) has been discussed recently but may not be feasible or cost effective for most vessels in the modern merchant fleet.

Marine vessels that carry cargo in bulk, such as oil tankers or dry bulk carriers, cannot generally avoid discharging ballast water in a cargo loading port. Part of the weight of the discharged bulk cargo, typically 50 percent, must be replaced with ballast water to maintain stability. However, there are other vessel types, such as passenger ships and container ships, which do not experience the same bulk shift in cargo that demands immediate ballast water replacement. These vessels provide opportunities for innovative designs and operational practices that can significantly reduce or even eliminate ballast water discharges in port.

Innovations to reduce ballast have also occurred in other types of vessels. Some vessels only require ballast water to replace fuel oil consumption. A recent research vessel design was able to use the processed effluent from the marine sanitation devices as ballast water. The mass balance between the crew's gray and black water waste was similar to the amount of ballast water required to account for consumed fuel oil. This approach eliminated traditional sea water ballast from the vessel design. The use of freshwater as ballast has also been proposed, either the onboard production of potable water as ballast for smaller vessels to replace fuel consumption or the transportation of freshwater from one port to another that might have limited supplies of drinking or agricultural water (e.g., Suban et al. 2010). Using a similar principle of only local water being onboard a vessel at any one time, other sorts of flow- through ballast systems have also been proposed. These approaches would likely require modifications to the existing ballast systems to actively and continuously pump water in and out of the ballast tanks throughout voyages, resulting in complete tank turnover in an hour or two.

Container ships can sometimes balance operations between loaded cargo and discharged cargo. Even when not balanced, the weight differential may often be within the margins of the vessel trim and stability requirements. One company has built and is operating two trailer-ships, similar to container ships that used design trades to eliminate the use of seawater ballast in all cases except emergencies. The ships are a bit wider, and potentially burn some additional fuel to account for their increased size. However, they have eliminated ballast water movements, as well as maintenance efforts associated with salt water piping systems and ballast tanks. Trim corrections are accounted for by shifting ballast water between tanks.

Given increased scrutiny and demands for ballast water exchange, it appears that many operators have been able to reduce or eliminate their discharges through careful operational practices, e.g., members of the Pacific Merchant Shipping Agency (PMSA) "all practice ballast water management protocols to reduce or eliminate the risk of introduction of aquatic invasive species in state waters ... . Over 80 percent of vessels hold all ballast water in port to eliminate this risk. Those vessels that must discharge ballast ensure that it is exchanged with mid-ocean water prior to entering coastal waters, dramatically reducing the risk of carrying invasive species" (Pacific Merchant Shipping Agency, National Environmental Coalition on Invasive Species, NECIS). Similarly, an industry led initiative, Marine Vessel Environmental Performance (MVeP), provides a numerical score to rate the environmental soundness of ballast water management. This score accounts for both the volume and the concentration of the ballast water discharged.

While many of these alternatives are conceptual at this point and may be limited to only specific vessels and/or routes, future ballast water management approaches to minimize the risk of invasive species may involve a variety of options and combination approaches. Regulatory frameworks for ballast water management that address both the volume and concentration of organisms in ballast discharges could further facilitate these alternative management approaches.

6.5.4. Temporal and Spatial Patterns

Independent of practices of ballast water uptake and total volume of a given discharge (previous sections), operational adjustments that modify the temporal and spatial patterns of ballast water discharge also may reduce the probability that discharged propagules will found a self-sustaining population (Drake et al. 2005). At least for sexually reproducing populations of planktonic species, for a given concentration of a given species in ballast discharge, the greater the volume discharged in a given time at a given location, the greater the probability of population establishment. If a total discharge volume for a given port of call can be broken up in space or time, invasion risk will be lowered. Thus, if a given discharge volume can be spread over space (e.g., as a vessel approaches harbor), be discontinuous in time (with scheduled breaks in discharge), or be discharged in a mixing environment (to dilute the concentration of propagules), the risk of invasion will be lowered (Drake et al. 2005).

For the same reasons, infrastructure modifications within ports that increase the rate and/or magnitude of dilution of discharged propagules also would decrease the risk of population establishment by discharged propagules. If discharges could be made in or piped to locations of greatest mixing within the harbor (e.g., closer to the tidal channels instead of in partially enclosed ship slips), then the rate of diffusion would be more likely to overcome the rate of reproduction. For example, low-velocity low- energy propellers, oloid mixers, or other mixing methods are routinely used in sewage treatment plants, industrial applications, and lakes. Such devices could be used in ports to increase the severity of Allee effects and other population hurdles faced by newly discharged propagules to minimize the probability of population establishment.

6.5.5. Combined Approaches

It may be possible to meet more stringent performance standards, or otherwise reduce the risk of invasions from ballast water discharges, by combining the approaches discussed in previous sections with either shipboard or onshore treatment. For example, a study by Fisheries and Oceans Canada suggests that conducting a mid-ocean exchange combined with BWMS for Great Lakes bound carriers may result in at least a 10x reduction in density of high risk taxa (Examining a combination treatment strategy: ballast water exchange PLUS treatment, Sarah Bailey, Fisheries and Oceans Canada). After considering the best science and technology now available, the state of Wisconsin is proposing to continue requiring ships to flush their ballast tanks at sea and require oceangoing ships to use BWMS to reduce remaining organisms to a level that meets the international numerical standard. This approach of combining ballast water exchange with shipboard ballast water treatment is targeting an enhanced level of protection for freshwater environments, similar to what has been proposed by Canada.

Each step from ballasting to deballasting, including the choice of procedures and the choice of technologies, contributes to the probability of an invasion occurring (see below). Recognizing and better quantifying the probability associated with each step could better target management efforts and achieve reductions in the overall probability of invasion at lower cost than relying only on BWMS.

6.6. Risk Management Approaches to Reduce Invasion Risk: Hazard Analysis and Critical Control Points (HAACP)

The Panel provides the following analysis as an example of one potential risk management approach that could be applied to ballast water management.

What Is HACCP?

Risk assessment for decision-making can be implemented using the Hazard Analysis and Critical Control Points (HACCP) approach. HACCP was developed in the late 1950s to assure adequate food quality for the nascent NASA program, further developed by the Pillsbury Corporation, and ultimately codified by the National Advisory Committee on Microbiological Criteria for Foods in 1997. The framework consists of a seven-step sequence:

1) Conduct a hazard analysis.
2) Determine the critical control points (CCPs).
3) Establish critical limit(s).
4) Establish a system to monitor control of the CCPs.
5) Establish the corrective action to be taken when monitoring indicates that a particular CCP is not under control.
6) Establish procedures for verification to confirm that the HACCP system is working effectively.
7) Establish documentation concerning all procedures and records appropriate to these principles and their application.

In international trade, these principles are important parts of the international food safety protection system. The development of HACCP ended reliance on the use of testing of the final product as the key determinant of quality, and instead emphasized the importance of understanding and control of each step in a processing system (Sperber and Stier 2009). HAACP principles also appear applicable to operationalize risk management for ballast water.

Basic Definitions

Hazard: The hazard under HACCP is the constituent whose risk one is attempting to control.

Critical control point: A critical control point (defined in the food sector) is “any point in the chain of food production from raw materials to finished product where the loss of control could result in unacceptable food safety risk” (Unnevehr and Jensen 1996).

Performance criteria: An important task in the HACCP process is to set performance criteria (critical limits) at each of the critical control points (CCP). The minimum performance criteria for each of the CCPs is set based on the final desired quality These criteria are determined using experimentation, computational models or a combination of such methods (Notermans et al. 1994). Then, readily measurable characteristics for each process needed to assure the desired quality are established and coupled to the control points.

Application in Food and Water

HACCP has been applied in the food safety area for 50 years, and in the past decade guidelines and regulations in the U.S. have been written that require an approved HACCP process in a number of applications. For example, the U.S. Food and Drug Administration has developed a HACCP process applicable to the fish and shellfish industries (21 CFR 123). HACCP has also been widely adopted in the EU, Canada and a number of other developed and developing nations for food safety (Ropkins and Beck 2000).

Havelaar (1994) was one of the first to note that the drinking water supply/treatment and distribution chain has a formal analogy to the food supply/processing/transport/sale chain, and therefore that HACCP would be applicable. However, in effect, the development of the U.S. surface water treatment rule under the Safe Drinking Water Act (40 CFR 141-142) and subsequent amendments incorporate a HACCP-like process. Under this framework, an implicitly acceptable level of viruses and protozoa in treated water was defined. Based on this, specific processes operated under certain conditions (e.g., filter effluent turbidity for granular filters) were "credited" with certain removal efficiencies, and a sufficient number of removal credits needed to be in place depending on an initial program of monitoring of the microbial quality of the supply itself. This approach (of a regulation by treatment technique) is chosen when it is not "economically or technically feasible to set an MCL [maximum contaminant level]" (Safe Drinking Water Act section 1412(b)(7)(A)).

How HACCP Might be Applied to Ballast Water Management

Shipboard BWMS and onshore treatment of ballast water differ in a number of characteristics that would affect their respective HACCP processes. Implementation of a HACCP program would need to account for different regulatory agencies and their scope of enforcement, the training of personnel, and the operational factors of each type of treatment. Figure 6-2 illustrates the control points for managing ballast water to reduce invasion risk; these are elaborated in the following examples of steps in applying the HAACP process:

- Identify the critical control points (which might include each particular treatment process as well as the method and type of intake water used)
- Determine the needed total reduction of organisms needed for the totality of the treatment system given the nature of the intake water (to achieve D-2, 10x D-2, etc.), and allocate these reductions amongst individual treatment processes.
- Given criteria in the discharged treated ballast water (D-2, 10x D-2, etc.), determine the minimum performance criteria for each treatment process, as well as criteria that determine whether or not particular intake water might be suitable. Note that these performance criteria should be based on easily measurable parameters that can be used for operational control. Research may be needed to determine relationships for each process between such surrogate parameters and removal of each of the size classes of organisms.

- A given ship having a set of processes with designated removal credits would only be allowed to take in ballast water that does not exceed the capacity of the controlled process train to meet the discharge criteria under the controlled operation.
- A QA process would be established for periodic validation and auditing (possibly by a third- party organization). Operational procedures would need to be developed to indicate the corrective actions needed for a particular process in the event that surrogate parameters fall outside acceptable limits, e.g., for additional holding time, recirculating for additional treatment, or some other measure.
- A blind testing procedure for the treatment products could be added to ensure that testing laboratory is not biased.
- Control points could also be identified for the various steps associated with transfer of an invasive species to a new habitat.

The overall context for applying HACCP methods varies depending upon the treatment envisioned. One criterion for deciding which option is more straightforward to implement may be the ease with which a risk management scheme can be applied and control ensured. In the consideration of shipboard and reception facilities, the Panel notes there are uncertainties in both approaches. The Panel has recommended that such uncertainties for ballast water treatment be assessed using a risk management framework such as HACCP. For this report, none of the BWMS or alternative ballast water approaches have been evaluated with respect to the risk of species invasions nor have critical control points been identified within the full sequence of ballast water management activities. Given the similarities among onshore treatment in reception facilities and water treatment facilities, some preliminary and partial extrapolations may be possible. However, the lack of information precludes further analysis of this issue by the Panel.

Figure 6-2 can be used as a heurist for identifying potential control points. For example, the characteristics of the port of origin could be included in the consideration of the types of propagules likely to be included in the ballast water. Known hazards from particular ports could be identified and the protocol for the control process modified for those ports. Open-ocean (or water) exchange is a way to reduce the number of propagules from the original port. Sea conditions or other factors may preclude an exchange. The control process may require modification to allow for this contingency. Next, there is transfer from the ballast tanks to the treatment system, which could be shipboard or on-shore. For either treatment, multiple control points could be identified. The role of sea chests, filter systems, oxidizing systems and plumbing could be identified. The HACCP approach would also take into account that onboard and on-shore treatment facilities will differ in the number and location of discharge points.

Likely outside of an engineering-based HACCP, but part of an overall strategy, is the consideration of the receiving waters for the ballast water and the types of habitat. Receiving habitats that are similar to those of the original port are likely to provide more opportunity for the establishment of an invasive species or pathogen. This information may be useful in establishing a site-specific treatment recommendation. These habitats could also be monitored as part of an overall plan for reducing the likelihood of successful invasion.

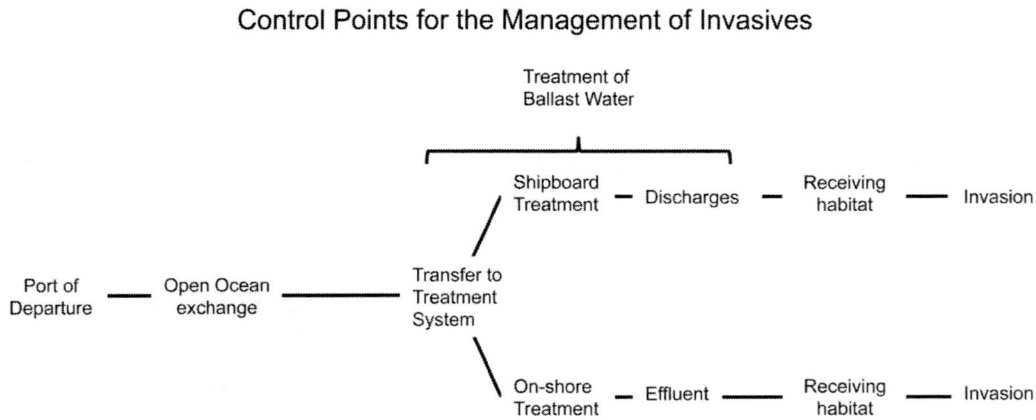

Figure 6-2. Some control points for the control of invasive species. Each of the processes may have imbedded control points.

HACCP can also be used to set priorities for the implementation of alternative means for managing ballast water. Most current BWMS are built with a one-size-fits-all approach and designed to be adopted by thousands of ships at some future time. There are defensible reasons for this one-size-fits-all approach, but as considered above, additional reasons exist to consider more flexible and combination approaches. This is especially true in the face of tight budgets and the constant need to prioritize spending on the most cost-effective strategies to reduce invasion risk. Setting priorities using HACCP principles could provide a basis for guiding the deployment of combinations of technologies and practices now and in the future (Keller et al. 2010). For example, to minimize invasion risk most cost- effectively while BWMS are being phased in, the highest risk ships that conduct the highest risk voyages could be retrofitted first. Likewise, ship-voyage specific risk assessments could guide the schedules for compliance monitoring of the operation and condition of installed water treatment systems.

6.7. Summary and Recommendations

6.7.1. Principal Limitations of Available Data and Protocols

- Data are not sufficiently compatible to compare rigorously across ballast water treatment systems because accepted standard protocols for testing ballast water treatment systems have been lacking, although they have been under development at multiple testing sites. The EPA ETV Protocol (U.S. EPA 2010) will improve this situation.
- No international requirement exists to report failures in type approval testing. On the basis of typically reported results, therefore, it is impossible to draw reliable conclusions about the consistency or reliability of some BWMS.
- The important size class of protists < 10 μm previously has been ignored in developing guidelines and standards.
- Clear definitions and direct methods to enumerate viable organisms in the specified size classes at low concentrations are missing for some size classes and indicator

organisms, and logistically problematic for all size classes, especially nonculturable bacteria, viruses, and resting stages of many other taxa.

6.7.2. Alternatives to Shipboard Treatment of Ballast Water

- Data on the effectiveness of practices and technologies other than shipboard ballast water treatment systems are inadequate because insufficient attention has been given to integrated sets of practices and technologies, including: (1) managing ballast uptake to reduce presence of invasives, (2) reducing invasion risk from ballast discharge through operational adjustments and changes in ship design to reduce or eliminate the need for ballast water, (3) development of voyage-based risk assessments and/or HACCP principles, and (4) options for treatment in reception facilities.
- Use of reception facilities for the treatment of ballast water appears to be technically feasible (given generations of successful water treatment and sewage treatment technologies), and is likely to be more reliable and more readily adaptable than shipboard treatment. Existing regional economic studies suggest that treating ballast water in reception facilities would be at least as economically feasible as shipboard treatment. However, these studies consider only vessels calling at those regional facilities; if vessels also call at ports outside the region without reception facilities, they would need a shipboard BWMS[11]. The effort and cost of monitoring and enforcement needed to achieve a given level of compliance is likely to be less for a smaller number of reception facilities compared to a larger number of BWMS.

6.7.3. Recommendations to Overcome Present Limitations

- Testing of BWMS in a research and development mode should be distinct from testing for certification, and certification testing should be conducted by a party independent from the manufacturer with appropriate established credentials, approved by EPA/USCG.
- Reported results from type approval testing of BWMS should include failures as well as successes during testing (as per the Protocol) so that the reliability of systems can be judged. This would be aided by the adoption of a transparent international standard format for reporting, including specification of QA/QC protocols.
- Consideration should be given to including protist-sized organisms < 10 μm in dimension in ballast water standards, and therefore in protocols to assess the performance of ballast water treatment systems.
- Consideration should be given to expanding test protocols recommended by the ETV to include components highlighted in Table 6-4.
- Suitable standard test organisms should be identified for bench-scale testing, and surrogate parameters should be investigated to complement or replace metrics that are logistically difficult or infeasible for estimating directly the concentration of living organisms.
- Use of representative "indicator" taxa (toxic strains of *Vibrio cholerae*; *Escherichia coli*; intestinal Enterococci) should continue as a sound approach to assess BWMS for effective removal of harmful bacteria. These estimates will be improved when

reliable techniques become available to account for active nonculturable cells as well as culturable cells.

- U.S. EPA is urged to develop metrics and methods appropriate for compliance monitoring and enforcement as soon as possible.
- Combinations of practices and technologies should be considered as potentially more effective approaches than reliance on one ballast water treatment technology. For example, ship-specific risk assessments (based on the environment and organisms present in previous ports of call) could be used to help prioritize the use of risk management practices and technologies, as well to target compliance and enforcement efforts.
- EPA should conduct a comprehensive analysis comparing biological effectiveness, cost, logistics, operations, and safety associated with both shipboard BWMS and reception facilities. If the analysis indicates that treatment at reception facilities is both economically and logistically feasible and is more effective than shipboard treatment systems, then it should be used as the basis for assessing the ability of available technologies to remove, kill, or inactivate living organisms to meet a given discharge standard. In other words, use of reception facilities may enable ballast water discharges to meet a stricter standard.
- Risk management is critical to ensure the efficacy of the entire spectrum of ballast water management; that is, not just the specific treatment process but also management practices, logistics and testing. Hazard Analysis and Critical Control Points (HACCP) has been demonstrated to be an effective risk management tool in a variety of situations and could be applied to ballast water management. HACCP methods are well understood and flexible. HACCP can be used to set priorities for ballast water management and can be applied to shipboard or shore-based systems or alternative management measures.

APPENDIX A. DOCUMENTS ON BALLAST WATER TECHNOLOGIES PROVIDED TO THE PANEL

This Appendix lists the documents available to the Panel for its assessment of ballast water technologies. These documents are available in the EPA Docket: Science Advisory Board Review of the Availability and Efficacy of Ballast Water Treatment Technology for EPA's Office of Water and the United States Coast Guard (at www.regulations.gov under docket number EPA-HQ-OW-2010-0582). Shaded rows indicate those documents that the Panel used as reliable sources of credible data for their assessment.

Table A-1. Documents Available to the Panel for its Assessment of Ballast Water Technologies

System	Document Title	Date
Group 1: Third-Party Reviews		
General	Ballast Water Treatment Technology: Current Status	2/1/2010
General	2009 Assessment of the Efficacy, Availability and Environmental Impacts of Ballast Water Treatment Systems for Use in California Waters	1/1/2009

Table A-1. (Continued)

System	Document Title	Date
General	October 2010 Update: Ballast Water Treatment Technologies for Use in California Waters	10/15/2009
General	Density Matters: Review of Approaches to Setting Organism-Based Ballast Water Discharge Standards	7/2/2005
General	International Convention for the Control and Management of Ships' Ballast Water and Sediments, 2004 - List of ballast water management systems that make use of Active Substances which received Basic and Final Approvals	9/24/2009
General	Ballast Water Treatment Advisory	6/8/2010
Group 2: Direct Data Reports and Supporting Information		
Ecochlor® Ballast Water Treatment System	STEP 2006 Application Form - Section 4.0: Proof of Ballast Water Treatment Performance	6/28/2005
Ecochlor® Ballast Water Treatment System	Final Environmental Assessment Review of the Application by Atlantic Container Lines for Acceptance of the Vessel M/V Atlantic Compass and the Ecochlor™ Inc. Technology into the USCG Shipboard Technology Evaluation (STEP) Program	8/1/2008
Ecochlor® Ballast Water Treatment System (Filtration +chlorine dioxide)	Final report of the land-based testing of the Ecochlor®-system, for Type Approval according to regulation-D2 and the relevant IMO guideline (April – July 2008)	2/1/2009
Electro-Clean Ballast Water Management System	Development of technologies on test facility and procedures for the land-based test as a type approval test at ballast water treatment system	6/30/2005
Electro-Cleen™ System	Information on the Type Approval Certificate of the Electro-Cleen™ System (ECS)	2/20/2009
GloEn-Patrol™ Ballast Water Management System	Type Approval Certificate of Ballast Water Management System	12/4/2009
Greenship's Ballast Water Management System	Landbased Test Report - Test Cycle Summary	6/29/2005
Hyde GUARDIAN Ballast Water Treatment System	Environmental Acceptability Evaluation of the Hyde GUARDIAN Ballast Water Treatment System as Part of the Type Approval Process	4/20/2009
Hyde GUARDIAN Ballast Water Treatment System (Filtration+UV)	Final report of the land-based testing of the Hyde-Guardian™ -System, for Type Approval according to the Regulation D-2 and the relevant IMO Guideline (April - July 2008)	1/1/2009
Hyde GUARDIAN Ballast Water Treatment System	Type Approval Certificate of Ballast Management System	4/29/2009
Hyde GUARDIAN Ballast Water Treatment System (Filtration+UV)	Shipboard Trials of Hyde "Guardian" System in Caribbean Sea and Western Pacific Ocean, April 5th - October 7th, 2008	4/1/2009
Hyde GUARDIAN Ballast Water Treatment System	Type Approval of the Hyde GUARDIAN™ Ballast Water Management System	5/7/2009
MSI (Filtration+UV)	MERC Land-Based Evaluations of the Maritime Solutions, Inc. Ballast Water Treatment System	11/1/2009
MH Systems (Deoxygenation)	Ballast water treatment by De-oxygenation with elevated CO2 for a shipboard installation	7/23/2003
NEI Venturi Oxygen Stripping (VOS)	Short-term Toxicity Testing of a De-oxygenation Ballast Water Treatment to Receiving Water Organisms. Final Report.	8/29/2008
NEI Venturi Oxygen Stripping (VOS)	Short-term Chronic Toxicity Testing of a De-oxygenation Ballast Water Treatment to Receiving Water Organisms. Final Report.	3/27/2009

System	Document Title	Date
NEI Venturi Oxygen Stripping (VOS)	STEP 2006 Application Form.	3/1/2006
NEI Venturi Oxygen Stripping (VOS)	Type Approval Certificate of Ballast Water Management System; Ballast Water Management System Type Approval Compliance Certificate	7/6/2009; 7/8/2007; 1/19/2010
NEI Venturi Oxygen Stripping (VOS) (deoxygenation & decavitation)	Application for Type Approval Certification: NEI Treatment Systems' Venturi Oxygen Stripping Ballast Water Management System.	3/1/2007
NEI Venturi Oxygen Stripping (VOS) (deoxygenation & decavitation)	Evaluations of a Ballast Water Treatment to Stop Invasive Species and Tank Corrosion.	6/27/2005
OceanSaver® Ballast Water Management System	Det Norske Veritas Type Approval Certificate	4/8/2009
OceanSaver® Ballast Water Management System	Type Approval Certificate of the OceanSaver ® BWMS	4/17/2009
OceanSaver® Ballast Water Management System	Information on the Type Approval Certificate of the OceanSaver® Ballast Water Management System	5/6/2009
OptiMarin Ballast System	Det Norske Veritas Type Approval Certificate	11/12/2009
Optimarin (Filtration+UV)	Land based testing of the OptiMarin ballast water management system of OptiMarin AS - Treatment Effect Studies	8/1/2008
Peraclean	Toxic Shock as New Ballast Water Treatment Fails Test	2/9/2010
PureBallast 250-2500	Det Norske Veritas Type Approval Certificate	6/27/2008
PureBallast (Filtration+UV+TiO2)	Land-based testing of the PureBallast Treatment System of AlfaWall AB	9/1/2008
PureBallast (Filtration+UV+TiO2)	Shipboard testing of the PureBallast Treatment System of AlfaWall AB	5/1/2008
SEDNA ® 250	Type Approval Certificate of Ballast Water Management System	8/16/2008
SEDNA® ballast water treatment system using PERACLEAN® Ocean	Effective Protection Against "Stowaways": Ballast Water Management System of Hamann and Evonik Receives Final Approval	6/11/2008
SEDNA®-System	Final report of the land-based and shipboard testing of the SEDNA®-system	3/1/2008
SEDNA®-System	Summary of Additional Provisions of the Type Approval Certificate of Ballast Water Management System SEDNA 250 of Hamann AG	8/1/2008
Severn Trent De Nora (BalPure)	Washington State Dept. of Fish and Wildlife Application Package Ballast Water Treatment System	8/8/2005
Severn Trent De Nora (BalPure)	Marrowstone Sodium Hypochlorite Mesocosm September 2004	9/1/2004
Severn Trent De Nora (BalPure)	Environmental Assessment Review of the Application for Acceptance of the SeaRiver Maritime Inc. S/R American Progress and Severn Trent de Nora BalPure™ System into the Shipboard Technology Evaluation Program (STEP)	2/1/2009
Severn Trent (Filtration+ electrochlorination)	Final report of the land-based testing of the BalPure®-BWT-System	1/1/2010
Severn Trent (Filtration+electrochlori nation)	MERC Land-Based Evaluations of the Severn Trent De Nora BalPure™ BP- 1000 Ballast Water Management System	7/1/2009
Siemens SICURE Ballast Water Management System	A Great Lake Relevancy Preamble to the GSI Report on Land-Based Testing Outcomes for the Siemens SICURE Ballast Water Management System	4/28/2010

Table A-1. (Continued)

System	Document Title	Date
GloEn-Patrol, Ecochlor, SiCURE, Resource Ballast Technologies System	Report of the tenth meeting of the GESAMP-Ballast Water Working Group (GESMP-BWWG)	10/30/2009
Siemens (Filtration+ electrochlorination	MERC Land-Based Evaluations of the Siemens Water Technologies SiCURE Ballast Water Management System	11/1/2009
Siemens SICURE Ballast Water Management System (Filtration+ electrochlorination)	Report of the Land-Based Freshwater Testing of the Siemens SiCURE Ballast Water Management System	5/15/2010
Group 3: G9 Files		
"ARA Ballast" Ballast Water Management System (formerly Blue Ocean Guardian BWMS)	Application for Final Approval of "ARA Ballast" Ballast Water Management System	3/23/2010
Alfa Laval Ballast Water Management System (PureBallast)	Basic Approval of Active Substances used by PureBallast management system	4/21/2006
Alfa Laval Ballast Water Management System (PureBallast)	Application for Final Approval of a ballast water management system using Active Substances	12/15/2006
AquaStar Ballast Water Management System	Application for Basic Approval of AquaStar Ballast Water Management System	3/18/2010
AquaTriComb Ballast™ Water Treatment System	Application for Basic Approval of the AquaTriComb Ballast Water Treatment System	12/16/2008
AquaTriComb Ballast™ Water Treatment System	Application for Basic Approval of the AquaTriComb™ Ballast Water Treatment System Corrigendum	6/29/2009
ATLAS-DANMARK, TG Ballastcleaner and TG Environmental guard, Sunrui Ballast Water Management System, DESMI Ocean Guard, Blue Ocean Guard (BOG)	Report of the eleventh meeting of the GESAMP-Ballast Water Working Group (GESMP-BWWG)	12/1/2009
BalClor TM ballast water management system (formerly Sunrui BWMS)	Application for Final Approval of BalClor TM ballast water management system	3/22/2010
BalPure®	Application for Basic Approval of the Severn Trent DeNora BalPure® Ballast Water Management System	8/28/2009
BalPure®	Application for Final Approval of the Severn Trent DeNora BalPure® Ballast Water Management System	3/28/2010
Blue Ocean Guardian (BOG) Ballast Water Management System	Application for Basic Approval of Blue Ocean Guardian (BOG) Ballast Water Management System	8/24/2009
Blue Ocean Shield Ballast Water Management System	Application for Basic Approval of the Blue Ocean Shield Ballast Water Management System	12/5/2008
BlueSeas Ballast Water Management System	Application for Basic Approval of the BlueSeas Ballast Water Management System	3/31/2010
CleanBallast!	Comments on the report of the fourth meeting of the GESAMP-BWWG	2/4/2008
CleanBallast!	Application for Final Approval of a ballast water management system using Active Substances	9/7/2007

System	Document Title	Date
CleanBallast!	Application for Final Approval of the RWO Ballast Water Management System (CleanBallast)	11/28/2008
ClearBallast, Greenship Sedinox, AquaTriComb	Report of the ninth meeting of the GESAMP-Ballast Water Working Group (GESMP-BWWG)	5/5/2009
DESMI Ocean Guard Ballast Water Management System	Application for Basic Approval of the DESMI Ocean Guard Ballast Water Management System	8/19/2009
EcoBallast	Application for Basic Approval of the HHI Ballast Water Management System (EcoBallast)	12/9/2008
EcoBallast	Application for Final Approval of HHI Ballast Water Management System "EcoBallast"	8/20/2009
Ecochlor® Ballast Water Treatment System	Application for Basic Approval of the Ecochlor® Ballast Water Treatment System	3/20/2008
Ecochlor® Ballast Water Treatment System	Application for Final Approval of the Ecochlor® Ballast Water Management System	12/16/2008
Ecochlor® Ballast Water Treatment System	Application for Final Approval of the Ecochlor® Ballast Water Management System	3/28/2010
EctoSys™	A Swedish Disinfection System	1/13/2006
EctoSys™	Basic Approval of Active Substances used by EctoSys™ electrochemical system	4/21/2006
Electo Clean System, Clear ballast System, CleanBallast! System	Report of the fourth meeting of the GESAMP-Ballast Water Working Group (GESAMP-BWWG)	12/19/2007
Electro-Clean Ballast Water Management System	Application for Basic Approval of Active Substances used by Electro-Clean (Electrolytic Disinfection) Ballast Water Management System	12/16/2005
Electro-Clean Ballast Water Management System	Application for Final Approval of a ballast water management system using Active Substances (Electro-Clean Electrolytic Disinfection)	9/7/2007
Electro-Clean Ballast Water Management System	Application for Final Approval of a ballast water management system using Active Substances (Electro-Clean Electrolytic Disinfection). Corrigendum	3/12/2008
Electro-Clean Ballast Water Management System	Application for Final Approval of the Electro-Clean System (ECS)	3/20/2008
En-Ballast	Application for Basic Approval of Kwang San Co., Ltd. (KS) Ballast Water Management System "En-Ballast"	8/25/2009
ERMA FIRST	Application for Basic Approval of the ERMA FIRST Ballast Water Management System	3/29/2010
General	Guidelines on the Installation of Ballast Water Treatment Systems	3/1/2010
GloEn-Patrol™ Ballast Water Management System	Basic Approval of Active Substance used by GloEn-Patrol™	9/7/2007
GloEn-Patrol™ Ballast Water Management System	Application for Final Approval of the GloEn-Patrol™ Ballast Water Treatment System	12/16/2008
Greenship Sedinox Ballast Water Management System	Application for Final Approval of the Greenship Sedinox Ballast Water Management System	12/12/2008
Greenship's Ballast Water Management System	Application for Basic Approval of a combined ballast water management system consisting of sediment removal and an electrolytic process using seawater to produce Active Substances (Greenship Ltd)	12/20/2007
HiBallast	Application for Basic Approval of Hyundai Heavy Industries Co., Ltd. (HHI) Ballast Water Management System (HiBallast)	8/24/2009
HiBallast, En-Ballast, OceanGuard, Severn Trent DeNora	Report of the twelfth meeting of the GESAMP-Ballast Water Working Group (GESMP-BWWG)	2/8/2010

Table A-1. (Continued)

System	Document Title	Date
Hitachi Ballast Water Purification System (ClearBallast)	Application for Basic Approval of Active Substances used by Hitachi Ballast Water Purification System (ClearBallast)	9/7/2007
Hitachi Ballast Water Purification System (ClearBallast)	Application for Final Approval of the Hitachi Ballast Water Purification System (ClearBallast)	12/11/2008
Hybrid Ballast Water Treatment System using Seawater Electrolytic Process	Basic Approval of Active Substances used by the Hybrid Ballast Water Treatment System using Seawater Electrolytic Process	12/14/2006
Hybrid Ballast Water Treatment System using Seawater Electrolytic Process, NKO3 BWTS, PureBallast, PureBallast	Report of the third meeting of the GESAMP-Ballast Water Working Group (GESMP-BWWG)	4/13/2007
Kuraray Ballast Water Management System	Application for Basic Approval of Kuraray Ballast Water Management System	3/25/2010
MES Ballast Water Management System (FineBallast MF)	Application for Basic Approval of the MES Ballast Water Management System (FineBallast MF)	3/17/2010
NK Ballast Water Treatment System	Request for re-evaluation of the proposal for the approval of Active Substances	8/18/2006
NK Ballast Water Treatment System	Basic Approval of Active Substances used by NK Ballast Water Treatment System	4/20/2006
NK-O3 BlueBallast System	Application for Final Approval of the NK-O3 BlueBallast System (Ozone)	3/21/2008
NK-O3 BlueBallast System	Application for Final Approval of the NK-O3 BlueBallast System (Ozone)	12/8/2008
OceanGuard™ Ballast Water Management System	Application for Basic Approval of the OceanGuard™ Ballast Water ManagementSystem	8/26/2009
OceanGuard™ Ballast Water Management System	Application for Final Approval of the OceanGuard™ Ballast Water ManagementSystem	3/25/2010
OceanSaver, Ecochlor, NK-O3 BlueBallast System	Report of the seventh meeting of the GESAMP-Ballast Water Working Group	7/28/2008
OceanSaver® Ballast Water Management System	Application for Basic Approval of a ballast water management system using Active Substances	9/7/2007
OceanSaver® Ballast Water Management System	Application for Final Approval of the OceanSaver® Ballast Water Management System (OS BWMS)	3/19/2008
Peraclean Ocean, ElectroClean	Report of the first meeting of the GESAMP-Ballast Water Working Group (GESMP-BWWG)	2/28/2006
Peraclean® Ocean	Application for approval of an Active Substance for Ballast Water Management	4/15/2005
Peraclean® Ocean	Application for approval of an Active Substance for Ballast Water Management. Corrigendum	5/27/2005
Peraclean® Ocean & Sedna system	Application for Final Approval of a ballast water management system using Active Substances	9/7/2007
Purimar™ Ballast Water Management System	Application for Basic Approval of Techwin Eco Co., Ltd. (TWECO) Ballast Water Management System (Purimar™)	3/9/2010
Resource Ballast Technologies System (cavitation combined with Ozone and Sodium Hypochlorite treatment)	Basic Approval of Active Substances used by Resource Ballast Technologies System (Cavitation combined with Ozone and Sodium Hypochlorite treatment)	4/6/2007

System	Document Title	Date
Resource Ballast Technologies System (cavitation combined with Ozone and Sodium Hypochlorite treatment)	Application for Final Approval of the Resource Ballast Technologies System (Cavitation combined with Ozone and Sodium Hypochlorite treatment)	12/19/2008
Resource Ballast Technologies System, GloEn Patrol, SEDNA using Percaclean Ocean, Ocean Saver	Report of the fifth meeting of the GESAMP-Ballast Water Working Group (GESMP-BWWG)	1/25/2008
Siemens SiCURE	Application for Basic Approval of the Siemens SiCURE Ballast Water Management System	12/19/2008
Special Pipe Ballast Water Management System (combined with Ozone treatment), NK Ballast Water Treatment System, EctoSys	Report of the second meeting of the GESAMP-Ballast Water Working Group (GESMP-BWWG)	7/7/2006
Special Pipe Hybrid Ballast Water Management System (with Ozone), CleanBallast, NK-O3 BlueBallast System, Blue Ocean Shield, EcoBallast	Report of the eighth meeting of the GESAMP-Ballast Water Working Group (GESMP-BWWG)	4/8/2009
Special Pipe Hybrid Ballast Water Management System combined with Ozone treatment version	Basic Approval of Active Substances used by Special Pipe Ballast Water Management System (combined with Ozone treatment)	4/12/2006
Special Pipe Hybrid Ballast Water Management System combined with Ozone treatment version	Application for Final Approval of the Special Pipe Hybrid Ballast Water Management System (combined with Ozone treatment)	12/4/2008
Special Pipe Hybrid Ballast Water Management System combined with Ozone treatment version	Application for Final Approval of the Special Pipe Hybrid Ballast Water Management System combined with Ozone treatment version (SP-Hybrid BWMS Ozone version)	3/17/2010
Special Pipe Hybrid Ballast Water Management System combined with PERACLEAN ® Ocean (SPO-SYSTEM)	Application for Final Approval of the Special Pipe Hybrid Ballast Water Management System combined with PERACLEAN ® Ocean (SPO-SYSTEM)	3/29/2010
Sunrui ballast water management system	Application for Basic Approval of Sunrui ballast water management system	8/24/2009
TG Ballastcleaner and TG Environmentalguard	Application for Basic Approval of the ballast water management system using "TG Ballastcleaner and TG Environmentalguard" as Active Substances (Toagosei Group)	12/26/2007
TG Ballastcleaner and TG Environmentalguard	Application for Final Approval of the JFE Ballast Water Management System (JFE-BWMS) that makes use of "TG Ballastcleaner® and TG Environmentalguard®"	8/20/2009
TG Ballastcleaner and TG Environmentalguard, Greenship's Ballast Water Management System, Electro-Clean System (ECS)	Report of the sixth meeting of the GESAMP-Ballast Water Working Group	7/14/2008

APPENDIX B. LITERATURE REVIEW OF RECEPTION FACILITY STUDIES

The Panel identified and evaluated a number of studies in the published and gray literature that discuss reception facilities (RF) (Table B-1).

Five studies have compared the effectiveness or costs of RF and shipboard treatment. In a study for the Canadian Coast Guard, Pollutech (1992) scored and ranked a variety of ballast water management approaches for vessels entering the Great Lakes, including ballast water exchange and several shipboard and land-based treatments, in terms of effectiveness, feasibility, maintenance and operations, environmental acceptability, cost, safety and monitoring. RF with discharge to a sanitary sewer (the only RF treatment scenario analyzed) ranked second out of 24 treatment and management approaches analyzed in the report.

In a second study for the Canadian Coast Guard, Aquatic Sciences (1996) considered RF alternatives (referred to as "pump off options") for Great Lakes shipping and found them to be "technically feasible" and to "undoubtedly offer the best assurance of prevention of unwanted introductions." The report further found that when installed onshore, "treatment options could have a more practical and enforceable application" than in shipboard installations, and concluded that "ship board treatment of ballast water appears to be logistically, economically, and particularly from the aspect of control, the least attractive method of ballast water treatment." The report estimated that treatment ships could be provided at key ports throughout the Great Lakes to receive discharged ballast water and heat it to >65°C at an annualized cost of around $17 to $51 million, or alternatively a single treatment ship could operate at a site en route to the Great Lakes to treat all incoming ballast water at an annualized cost of $2.7-2.8 million. Retrofitting costs to enable ships to discharge their ballast water to treatment ships were estimated at approximately $40,000 to over $200,000 per ship.

AQIS (1993a) developed designs and cost estimates to compare shipboard, land-based and treatment ship-based treatment at a port serving 140,000-ton bulk carriers. The shipboard design consisted of a 50-µm strainer, with high-level ballast tank off-take pipes to reduce the discharge of ballast sediments and settled cysts or spore stages. The land-based designs included either 4,000 or 52,000 MT of storage, with coagulation, flocculation, granular filtration, UV disinfection, and thickening, dewatering and disposal of solids. The treatment ship design included 4,000 MT of storage, pressurized granular filters, UV, and solids management and disposal. Annualized costs were reported as $0.69/MT for shipboard treatment, $0.55/MT for treatment in a treatment ship, and $0.35-$0.62 for treatment in a land-based facility (depending on the type and size of storage used).[12] Some costs (pipelines to transport ballast water from berths to treatment plants, and land costs) were not included in the RF alternatives, which reduced their estimated cost relative to the shipboard alternative. On the other hand, the Panel notes that the RF treatment analyzed here (granular filtration with coagulation and flocculation, followed by UV disinfection) would treat ballast water to a substantially higher standard than the shipboard alternative (a 50 µm strainer with no disinfection); and that basing the analysis on large bulk carriers, which typically discharge the largest volumes of ballast water of the vessels using Australia's ports (Table 4.1 in AQIS 1993a), favored shipboard treatment.

Table B-1. Reports that discuss reception facilities (RFs) for onshore treatment of ballast water

Report	Discussion	Conclusions
Pollutech 1992	Compares and ranks various shipboard and RF treatment approaches.	RF ranks 2nd out of 24 options.
AQIS 1993a	Compares shipboard, land-based and treatment ship approaches in Australia.	Land-based and treatment ship are cheaper and more effective than shipboard.
AQIS 1993b	Briefly discusses treatment ship and land-based treatment in Australia.	RF is unlikely except in special circumstances.
Ogilvie 1995	Reviews possible treatment methods and estimates some costs for RFs.	Several methods show promise for RF.
Aquatic Sciences 1996	Compares shipboard, treatment ship, land-based and external source treatment.	RF is technically feasible and the most effective and cheapest approach.
NRC 1996	Briefly discusses advantages and disadvantages of RF.	RF remains an option.
Gauthier & Steel 1996	Mentions shipboard, treatment ship and land-based approaches.	RF is considered a poor option.
Victoria ENRC 1997	Briefly discusses RFs.	RF is probably too costly at a large scale; may be viable at a smaller scale.
Greenman *et al.* 1997	Student report commissioned by the U.S. Coast Guard.	RF is feasible at all sites considered.
Cohen 1998	Briefly discusses advantages and disadvantages of RFs.	
Reeves 1998, 1999	Briefly discusses RFs.	Lists RF as an alternative.
Oemke 1999	Briefly discusses advantages and disadvantages of RF.	RF is feasible for some parts of the industry, such as VLCCs.
Dames & Moore 1998, 1999	Briefly discusses RF.	RF may be good option at oil export terminals with oil stripping plants.
Cohen & Foster 2000	Briefly discusses advantages and disadvantages of RF.	
CAPA 2000	EPA-funded study estimates the cost of RF for California.	RF is technically feasible.
Rigby & Taylor 2001a,b	Briefly discusses RF.	Cost, availability, quality control may prevent RF development, but it might work for tankers that discharge oily ballast to RFs.
US EPA 2001	Briefly mentions RF.	
California SWRCB 2002	Briefly discusses RF.	RF is an attractive option, at least for some parts of the industry.
Glosten 2002	Estimates upper-bound retrofit costs to discharge ballast to RFs.	
NSF 2003	Mentions shipboard, RF and operational options for the longer term.	Shipboard seems the most challenging approach.
Hilliard 2006; Hilliard & Matheickal 2010	Compares and ranks various shipboard and RF treatment approaches for the Black Sea-Caspian Sea Waterway.	RF ranks 1st out of 16 options.
Brown & Caldwell 2007, 2008	Develops designs and estimates costs for RF at the Port of Milwaukee.	RF is feasible; treatment ship is cheaper than land- based.
California SLC 2009, 2010	Briefly discusses advantages and disadvantages of RF.	RF might be suitable for terminals with regular vessel calls such as cruise ships, or for the Port of Milwaukee.
Pereira *et al.* 2010	Uses simulation model to assess RF operation at a Brazilian port.	RF treatment would not affect port operations negatively.
Donner 2010a,b,c	Compares RF and shipboard treatment.	RF is more efficient, cheaper and safer.

AQIS (1993a) also developed a scenario for RF treatment of all the ballast water discharged in Australia, using both treatment ships and land-based treatment plants. Total capital costs for RF treatment were estimated at $330 million and annual operating costs at

$6.7 million. The capital cost for outfitting one-year's worth of visiting ships for shipboard treatment was estimated at $1 billion "ignoring the fit out of new ships in future years," with estimated annual operating costs working out to $5.4 million. The study concluded that "land-based or port-based [=treatment ship] facilities are more economic and effective than numerous ship-board plants."

California's State Water Resources Control Board (California SWRCB 2002) qualitatively evaluated RFs and ten shipboard treatment alternatives for effectiveness, safety, and environmental acceptability. RF was the only approach rated acceptable in all three categories. There were reservations or unresolved questions about the effectiveness of all shipboard alternatives, about the safety of 80 percent of the shipboard alternatives, and about the environmental acceptability of 90 percent of the shipboard alternatives.

Hilliard (2006) (also reported in Hilliard and Matheickal 2010) compared an RF using conventional water treatment methods (such as granular filtration with disinfection) to 15 shipboard treatment approaches for vessels transiting the Black Sea-Caspian Sea Waterway. Based on scores for 13 technical factors, RF treatment was ranked first. The study concluded that an RF, using standard water industry methods, would provide a cost effective solution if based at an appropriate port, but cautioned that this might be a less useful approach for some vessels.

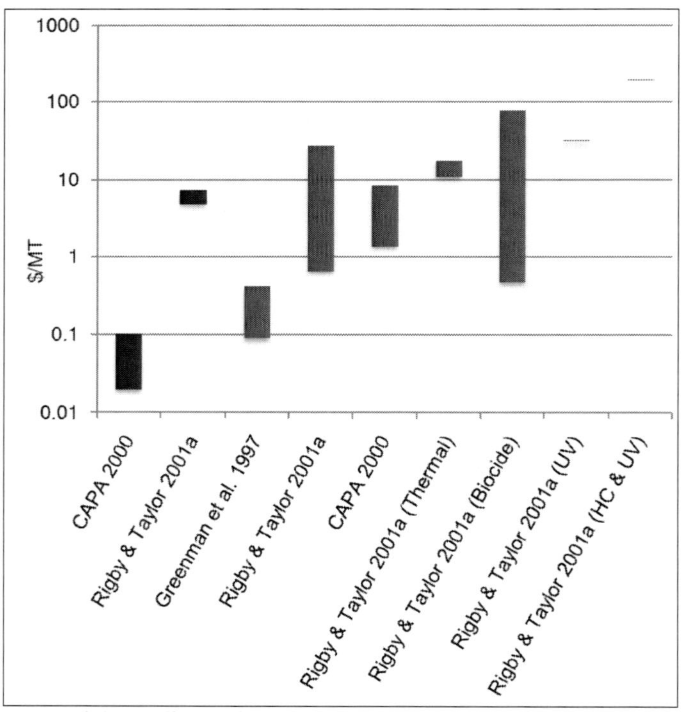

The Coast Guard converted Australian estimates to U.S. dollars at the Oct. 16, 2001 exchange rate, but did not adjust estimates for inflation. One m3 of ballast water is assumed to weigh 1 MT. Cost estimates (note log scale) for ballast water exchange are in blue, for RF treatment in green, and for shipboard treatment in red.

Figure B-1. Cost estimates listed in U.S. Coast Guard (2002).

In each of these comparative studies, RF was judged to be as effective or more effective, and generally cheaper, than shipboard treatment. As noted, there are limitations to these studies and grounds for criticism; in particular, some were done over a decade ago and do not reflect current BWMS costs. However, these studies comprise the most detailed published comparisons of RF and shipboard treatment approaches available. In addition, the U.S. Coast Guard compiled a table of cost estimates from different studies (U.S. Coast Guard 2002). Figure B-1 shows all the estimates that were expressed as costs per metric ton or cubic meter of ballast water, and thus in a form that can be compared. In these estimates, RF treatment is generally more expensive than ballast water exchange and less expensive than shipboard treatment, though there is considerable overlap. These cost estimates also do not reflect current BWMS costs.

In three recent papers, Donner (2010a,b,c) argued that RF treatment is more efficient, less expensive, safer for the crew and the environment, and easier to monitor and verify than shipboard treatment, but that shipboard treatment has been pursued because it is "the solution of least resistance" politically. Other comparisons of RF and shipboard treatment in the literature consist of lists or brief discussions of their relative merits. These reports variously concluded that RF treatment is probably a superior or probably an inferior option compared to shipboard treatment, or that RF treatment is suitable for a particular part of the cargo fleet (Table B-1), but none provided analysis or data to support these conclusions.

Two studies (in addition to AQIS (1993a) and Aquatic Sciences (1996), discussed above) provided conceptual designs and cost estimates for RF treatment for specific regions. CAPA (2000), an EPA- funded study conducted for the California Association of Port Authorities, developed conceptual designs and cost estimates for constructing and operating ballast water treatment plants at cargo ports in California (Table B-2). These plans and estimates included pipelines from berths to plants; storage tanks; coagulation, flocculation, filtration and UV disinfection; thickening, dewatering and landfill disposal of residual solids; and discharge of effluent through an outfall pipeline. They did not estimate costs for land, permits, seismic evaluation, or retrofitting vessels to enable them to discharge ballast water to an RF. The study concluded that onshore treatment would be technically and operationally feasible, though there could be delays to vessels in some circumstances.

Brown & Caldwell (2007, 2008) developed designs and cost estimates for land-based and treatment ship approaches to treat ballast discharges from oceangoing ships arriving at the Port of Milwaukee. The first report assessed four land-based treatment systems:

- 100-μm screening followed by UV treatment;
- Coarse screening followed by ozonation;
- 500-μm screening followed by membrane filtration to remove particles > 0.1 μm;
- 500-μm screening followed by hydrodynamic cavitation.

These were analyzed along with two systems for transferring and storing the discharged ballast water: discharge at berths into pipelines to land-based RF with storage tanks; and discharge to a barge that would store and carry the water to a land-based RF. Design criteria required a system capable of receiving ballast water at 680 MT/h, storage capacity of 1,900 MT, and treatment at 80 MT/h. The report concluded that all four treatment systems and both transport/storage systems are feasible, with UV treatment and hydrodynamic cavitation having the most promise for treating viruses (Brown & Caldwell 2007). The second report

(Brown & Caldwell 2008) developed a design and cost estimate for retrofitting a barge to serve as a treatment ship, which would collect, store and treat ballast water. Treatment included a cloth media disk filter with a nominal pore size of 10 µm, and UV disinfection at an estimated minimum dose of 30 mJ/cm^2. The design criteria for this analysis included the capacity to receive ballast discharges at 2,300 MT/h, storage of 10,000 MT, and treatment at 230 MT/h, which is around three times the flow rates and five times the storage required in the first report. Estimated costs from both studies are shown in Table B-3.

Table B-2. Cost estimates for onshore treatment in California

System Component	Capital Costs	Annual O&M
Pipelines	146,950,000	–
Storage Tanks	76,235,000	–
Treatment Plants	22,510,000	2,018,000
Outfalls	1,380,000	–
Total	247,075,000	2,018,000

Source: CAPA 2000.

Table B-3. Cost estimates for onshore treatment for oceangoing ships at the Port of Milwaukee

Treatment [1]	Transport	— — Capital Costs — —				Annual O&M
		Pipelines[4]	Storage	Treatment	Total	
100-µm screening & UV [1]	Pipelines	2,973,000	1,252,000	615,000	4,840,000	11,500
Ozone [1]	Pipelines	2,973,000	1,252,000	835,000	5,060,000	9,800
0.1-µm membrane filter [1]	Pipelines	2,973,000	1,252,000	1,096,000	5,321,000	15,600
Hydrodynamic cavitation [1]	Pipelines	2,973,000	1,252,000	2,608,000	6,833,000	20,900
100-µm screening & UV [1,2]	Barge	261,000	522,000	615,000	1,398,000	367,000
Ozone [1,2]	Barge	261,000	522,000	835,000	1,617,000	365,000
0.1-µm membrane filter [1,2]	Barge	261,000	522,000	1,096,000	1,879,000	371,000
Hydrodynamic cavitation [1,2]	Barge	261,000	522,000	2,608,000	3,390,000	376,000
10-µm filter & UV [3]	Treatment ship	0	2,695,000	866,000	3,561,000	514,000

Source: Brown & Caldwell 2007, 2008.

[1] Design criteria are maximum ballast discharge of 680 MT/h,1,900 MT storage, and treatment rate of 80 MT/h.

[2] "Storage" refers to barge purchase and modification to use for ballast water transfer and storage, exclusive of the treatment system.

[3] Design criteria are maximum ballast discharge of 2,300 MT/h, 10,000 MT storage, and treatment rate of 230 MT/h.

[4] Includes collection pumps, pipelines, a lift station and coarse screening.

Besides the need for facilities to receive, store and treat ballast water from ships, ships must be modified so they can safely and rapidly discharge ballast water to RFs. This requires modification of a ship's pipe system and possibly larger ballast pumps, to raise the water to deck level and/or to discharge it quickly enough. Cost estimates have ranged from around $15,000 to $540,000 for container ships (Pollutech 1992; Glosten 2002), $15,000 to $500,000 for bulkers (Pollutech 1992; CAPA 2000), and less than $140,000 to around $2.3 million for tankers (Victoria ENRC 1997; Glosten 2002) (Table B-4). Most of these estimates explicitly included the replacement of existing pumps with more powerful pumps where needed (AQIS 1993a; Aquatic Sciences 1996; Dames & Moore 1998; CAPA 2000; Glosten 2002; Brown & Caldwell 2008[13]). The cost to outfit a new ship was estimated to be less than the cost to retrofit an existing ship (AQIS 1993b), perhaps by an order of magnitude (CAPA 2000). Some reports provided little or no explanation of their retrofit/modification estimates (Pollutech 1992; AQIS 1993a; Aquatic Sciences 1996; Dames & Moore 1998). Victoria ENRC (1997) provided a materials list for a bulk carrier, and noted that a tanker "with its ballast lines running on deck would have a considerable lower installation cost." CAPA (2000) provided a cost breakdown for modifying a bulker, and stated that modifying a tanker would generally cost more.

Table B-4. Cost estimates for retrofitting ships to discharge ballast water to a treatment facility

Ship Type	Capital Cost	Report
Great Lakes bulker, break-bulk or container	$13,200–26,500	Pollutech 1992
Small container	$20,400	AQIS 1993a
Large bulker (140,000 DWT; 45,000 MT; 4,000 MT/h)	$204,000	AQIS 1993a
Great Lakes bulker	$40,400–202,00	Aquatic Sciences 1996
Handysize bulker (520'; 22,000 DWT)	$142,000	Victoria ENRC 1997
Container	$53,200-173,000	Dames & Moore 1998 [1]
Container or bulker (1,000 MT/h)	$502,000	CAPA 2000
Tanker (869'; 123,000 DWT; 75,850 MT; 6,400 MT/h)	$2,3230,000	Glosten 2002
Bulker (735'; 67,550 DWT; 35,000 MT; 2,600 MT/h)	$131,000	Glosten 2002
Break-bulk (644'; 40,300 DWT; 26,850 MT; 3,000 MT/h)	$373,000	Glosten 2002
Container (906'; 65,480 DWT; 19,670 MT; 2,000 MT/h)	$540,000	Glosten 2002
Car carrier (570'; 13,847 DWT; 6,600 MT; 550 MT/h)	$198,000	Glosten 2002
Bulker (469'; 5,700 MT; 570 MT/h)	$60,000	Brown & Caldwell 2008
Bulker (722'; 18,000 MT; 2,300 MT/h)	$203,000	Brown & Caldwell 2008

Where the data are available, length is given in feet, size in deadweight tons (DWT), ballast water capacity in metric tons (MT), and maximum ballast discharge rate in metric tons per hour (MT/h), in parentheses following the ship type.

[1] Estimate developed by the Pacific Merchant Shipping Association.

Glosten (2002) and Brown & Caldwell (2008) provided the most detailed estimates. Glosten (2002) estimated ship retrofit/modification costs for five ships representing common vessel types in Puget Sound (Table B-4). The modifications were designed to “allow ballast transfer with minimal disruption to current operations” including sizing them for deballasting completely at berth during the time needed to load cargo, eliminating any need to start deballasting before arriving at berth. For each vessel type, the authors selected ships that “had ballast systems with capacities on the upper end of vessels that call on Puget Sound to attempt to establish an upper-bound on retrofitting costs.” In selecting pipe sizes and other elements “every attempt was made to capture an upper bound on the modification costs associated with each vessel type surveyed,” including the installation of “a completely new piping system to provide the ability to fill and empty each ballast tank separately.” Notably, this new piping system was included in the tanker estimate even though it is not needed on crude oil tankers, the type of tanker analyzed, where “a simpler, lower-cost solution” exists. It was included because it could be needed on some other ships (i.e. product tankers) in the same general category, and this produced by far the highest cost estimate in the study.[14] The modifications were also designed to allow ballast water transfer in either direction between a ship and a RF (either onto or off a ship),[15] which in some cases may raise the cost over what is needed to only discharge ballast water to RFs.

Brown & Caldwell (2008) provided analyses, conceptual designs, drawings and cost estimates for modifying two sizes of ocean-going bulkers serving the Great Lakes, based on a smaller actual ship and a larger hypothetical ship (Table B-4). These designs were also sized to allow the ship to initiate and complete deballasting at berth during cargo loading.

The potential for treating ballast discharges with RFs has also been recognized in laws, regulations, guidelines and treaty conventions. The U.S. Nonindigenous Aquatic Nuisance Prevention and Control Act (NANPCA) of 1990 and the National Invasive Species Act (NISA) of 1996 directed the U.S. Coast Guard to fund research on ballast water management, specifically noting that technologies in “land- based ballast water treatment facilities” could be included, and to investigate the feasibility of using or modifying onshore ballast water treatment facilities used by Alaskan oil tankers to reduce the introduction of exotic organisms (§§1101(k)(3), 1104(a)(1)(B), 1104(a)(2) and 1104(b)(3)(A)(ii) in U.S. Congress 1990, 1996). In the interim and final rules implementing NISA, the U.S. Coast Guard specifically included discharge to a RF as a means of meeting NISA’s ballast discharge requirements, and required ships to keep records of ballast water discharged to RFs (US Coast Guard 1999, 2001), although the Coast Guard eliminated these provisions when it concluded that it did not have the authority to regulate or approve RFs (US Coast Guard 2004). The U.N. International Maritime Organization’s 1991 Guidelines stated that “Where adequate shore reception facilities exist, discharge of ship’s ballast water in port into such facilities may provide an acceptable means of control” (IMO 1991 and IMO 1993, §7.5 Shore Reception Facilities). The IMO’s 1997 Guidelines stated that “Discharge of ship’s ballast water into port reception and/or treatment facilities may provide an acceptable means of control. Port State authorities wishing to utilize this strategy should ensure that the facilities are adequate...If reception facilities for ballast water and/or sediments are provided by a port State, they should, where appropriate, be utilized” (IMO 1997, §7.2.2, §9.2.3). The IMO’s 2004 Convention stated that “The requirements of this regulation do not apply to ships that discharge ballast water to a reception facility designed taking into account the Guidelines developed by the Organization for such facilities” (IMO 2004, Regulation B-3.6). The IMO adopted specific guidelines for

RFs (IMO 2006), and recognized RFs as an alternative in IMO 2005b (§1.2.3), as have Australia, New Zealand and Canada in their ballast water regulations (AQIS 1992; New Zealand 1998, 2005; Canada 2000, 2007).

APPENDIX C. FURTHER INFORMATION ON STATISTICS AND INTERPRETATION

A major challenge of sampling at low organism concentrations is that many samples will have zero live organisms because the few live organisms present are missed. Therefore to improve the probability of detecting them, large volumes must be sampled and excellent techniques must be used to enable detection (Figure C-1).

Consider the following examples from Lee et al. (2010): from the Poisson distribution, if 1 m^3 of ballast water was sampled from a ballast water discharge that had a known average concentration of 10 zooplankton-sized organisms m^{-3} over the entire ballast water volume, about 95 percent of the samples would contain 4-17 organisms m^{-3}. As the concentration of organisms decreases, the frequency distribution becomes increasingly skewed, and there is a high probability of obtaining a sample with zero organisms. Thus, if the sample concentration is 1 organism m^{-3}, the probability of a 1 m^3 sample containing zero organisms is 36.8 percent. If the sample concentration is only 0.01 organism m^{-3}, or 1 organism in 100 cubic meters of ballast water, the probability of obtaining a sample with zero organisms is ~99 percent. Moreover,

> If a small volume is used to evaluate whether the discharge meets a standard, the sample may contain zero detectable organisms, but the true concentration of organisms may be quite high....For example, even with a relatively high concentration of 100 organisms m^{-3}, only about 10% of 1-L samples will contain one or more organisms. Furthermore, even if zero organisms are detected in a 1-L sample, the upper possible concentration, based on a 95% confidence interval, is about 3,000 organisms m^{-3}....The general point is that more organisms may be released in ballast discharge using a stringent standard paired with a poor sampling protocol than a more lenient standard paired with a stringent sampling protocol. (Lee et al. 2010, p.72).

The available methodologies for testing compliance with the IMO standards for zooplankton-sized organisms are at or near the analytic detection limits. The following example from the ETV Protocol (U.S. EPA 2010) illustrates the problem: For the desired minimum precision in quantifying zooplankton-sized organisms, consider an example where the upper bound of the Chi-square statistic should not exceed twice the observed mean (corresponding to a coefficient of variation of 40 percent, which is relatively high). Then, ***if 6 or fewer*** live organisms are counted, the upper bound of the 95% CI for the volume sampled does not exceed the IMO/Phase-1 performance standard for zooplankton-sized organisms (< 10 viable individuals per m^3):

- Coefficient of variation (CV) = standard deviation (SD) divided by the mean (M).
- For the Poisson distribution, the variance (V) = SD^2 = M.
- Substituting the critical value of the mean, 6: CV = $6^{1/2}/6 \approx 40\%$.

The volume needed to find and quantify 6 live organisms per m^3 depends on the whole-water sample volume, the concentration factor, and the number of subsamples examined. Very large sample volumes (tens of m^3) are required to quantify viable zooplankton-sized organisms (assuming 20 mL of the concentrated sample is analyzed), and each sample must be concentrated down to a manageable volume (concentrating 3 m^3 to 1 L would yield a concentration factor of 3,000). Based on the Poisson distribution for a 95% CI from the Chi-square distribution, 30 m^3 (30,000 L) must be sampled in order to find and count < 10 organisms m^{-3} with the desired level of precision. The total sample volume can be reduced if the concentration factor is increased (and the same subsample volume analyzed), if the CI is also lowered (e.g., from 95% to 90%) or the subsample volume analyzed is increased (e.g., from 20 mL to 40 mL). Notably, as the concentration factor increases, the likelihood of losing organisms or inadvertently killing them in the sample processing steps increases, thus creating an artifact that overstates the effectiveness of the treatment (see Section 6.2.4 (A)).

The ETV Protocol provides examples of the sample size needed to provide the level of precision needed to achieve a 95% upper confidence limit that is no more than twice the observed mean and does not exceed the targeted concentration. If the volume of subsample that is analyzed is increased, then validation experiments should be conducted to ensure that counting accuracy is acceptably high. The problem is exacerbated for zooplankton-sized organisms because they are sparse compared to organisms in the next smaller class (here, referred to as "protist-sized" organisms, or organisms ≥10 μm and < 50 μm in minimum dimension). The Poisson distribution assumption still applies to this smaller size class, and the ETV Protocol provides examples with a more stringent level of precision than is used for the larger size class (Table C-2; U.S. EPA 2010). At present, confirmation of the Phase 1 standard (< 10 protist-sized organisms mL^{-1}) represents the practical limit that can currently be achieved by testing facilities in the U.S. (e.g., MERC 2009, 2010a, 2010b; Great Ships Initiative 2010).

Laboratory Experiments with Protist Cultures Support Use of the Poisson Distribution

A workshop was held to evaluate four methods for enumerating living protists in treated ballast water (Nelson et al. 2009, Steinberg et al., accepted with revisions). Live and dead cells were counted using flow cytometry, an enhanced flow-through system with imaging capacity (FlowCAM®, Fluid Imaging Technologies, Yarmouth, ME), direct counts of samples collected on membrane filters, and direct counts using a Sedgewick Rafter counting chamber. All techniques used fluorescent stains to differentiate between live and dead cells. Counting methods were tested with several ratios and densities of live and dead *Tetraselmis* sp., a small phytoflagellate. In these trials, comparisons were conducted under ideal conditions with no debris (except for one sample) or particulate matter and with a single target species.

Data were evaluated to determine whether they conformed to a Poisson distribution by determining if the variance was equal to the mean. At low concentrations of living cells (approximately 10 mL^{-1} to 100 mL^{-1}), there was no evidence to reject the Poisson hypothesis (Nelson et al. 2009).

Table C-1. Sample volume of treated ballast water required relative to performance standards for organisms ≥10 μm and <50 μm (nominally protists), assuming that the desired level of precision is set at a CV of < 10%. These are the required whole-water sample volumes that must be concentrated to 1 L as a function of N, the number of 1-mL subsamples analyzed

N=	2	3	4
Concentration (i.e. performance standard) (individuals mL-1)	Sample Volume Required (L)		
0.01	6,000	4,000	3,000
0.1	600	400	300
1	60	40	30
10	6	4	3

Source: U.S. EPA 2010.

Accuracy and Precision in Sparse Samples Following a Poisson Distribution

A series of laboratory experiments was conducted to assess the accuracy and precision of enumerating zooplankton- and protist-sized organisms at a variety of densities (Lemieux et al. 2008). Inert 10-μm standardized microbeads at densities of 1, 5, 10, 50, 100, 500, and 1,000 microbeads per mL of artificial seawater represented protist-sized organisms, and 150-μm microbeads at 10, 30, and 60 microbeads per 500 mL represented zooplankton-sized organisms. Such inert, standardized polymer microbeads were used rather than organisms to eliminate any potential bias, and artificial (rather than natural) seawater was used to avoid inclusion of various organic particles (e.g., detritus) that could interact with the microbeads and confound interpretations. Here, microbeads served as proxies, and it is acknowledged they are an imperfect representation of living organisms (e.g., microbeads do not exhibit the swimming behavior of many planktonic organisms; living organisms can cling to nets or filters and can also be squeezed through a net more readily than microbeads). Nonetheless, if the items of interest in a sparse concentration – be they microbeads or living organisms in treated ballast water – are well mixed, the Poisson distribution should be applicable. In addition, as stated previously, a continuously isokinetically taken sample whose contents are completely counted avoids problems associated with the spatial distribution of the organisms.

At each microbead density, the percent difference of the observed mean from the expected mean indicated counting accuracy, and the CV indicated the level of precision. In this study, benchmarks for acceptable accuracy and precision were established at a percent difference of 10 percent and a CV of 0.2 (20 percent), respectively. For the "protist" microbeads, the 50 - 1,000 mL^{-1} concentrations were not significantly different, with acceptable accuracy and precision below the 10 percent and 20 percent benchmarks, respectively. Unfortunately, however, analysis of the "zooplankton" microbead populations at all densities showed poor precision, with CVs well above 20 percent, and only counts at the highest density showed a CV < 100 percent. All densities of "zooplankton" microbeads showed acceptable accuracy (i.e., a percent difference < 10%) after sufficient aliquots were examined to result in a stable mean.

From this work, Lemieux et al. (2008) recommended that samples for analysis of protist-sized organisms should be concentrated by at least a factor of five, and that at least four replicate counting chambers (e.g., four Sedgewick Rafter slides) should be analyzed for acceptable accuracy and precision, including evaluation of at least 10 random rows (from a total of 20) of each counting chamber. Importantly, for zooplankton-sized organisms, Lemieux et al. (2008) determined the earlier (draft) ETV protocol recommendations for sample sizes as inadequate to achieve acceptable precision. The data from these microbead experiments indicated, instead, that this size class requires a sample size of greater than 6 m^3, concentrated to 0.5 L (i.e., concentrated by a factor of 12,000), and analysis of at least 450 1-mL aliquots, considering that CVs at the highest volumes were > 20%. As higher concentration factors were likely unrealistic, it was suggested that larger sample sizes and improved analytical methods should be used. Lemieux et al. (2008) also noted that these laboratory trials represented a "best case" situation because the study was conducted under simplified, "ideal" conditions rather than with natural organism assemblages in natural seawater.

When concentrations are close to the performance standard, a single sample may require too large a volume of water to be logistically feasible. In that case, complete, continuous time-integrated sampling (with the entire volume analyzed) and combining samples across multiple trials can improve resolution while maintaining statistical validity. To that end, Miller et al. (2011) applied statistical modeling (based on the Poisson distribution) to a range of sample volumes and plankton concentrations. They calculated the statistical power of various sample volume and zooplankton concentration combinations to differentiate various zooplankton concentrations from the proposed standard of < 10 m^{-3}. Their study involved a two-stage sampling approach. Stage 1 checked compliance based on a single sample, which was expected to be effective when the degree of noncompliance was large. Stage 2 combined several samples to improve discrimination (1) when concentrations are close to the performance standard, or (2) when a large-volume, single-trial sample would be logistically problematic, or both. The Stage 2 approach took advantage of the fact that the sum of several Poisson random variables is still a Poisson distribution, and is called the "summed Poisson method." Stage 2 also compared the summed Poisson approach to power calculations using standard t-tests, the nonparametric Wilcoxon Signed Rank test (WSRT), and a binomial test, all well-known statistical techniques. The summed Poisson approach had more statistical power relative to the other three statistical methods. Not surprisingly, as noncompliant concentrations approached the performance standard, the sampling effort required to detect differences in concentration increased. The major conclusions from this study are presented in Section 3.2.1.

The major finding from Miller et al. (2011) is that three trials of time-integrated sampling of 7 m^3 (and analyzing the entire concentrated sample from the 21 m^3) from a ship's BW discharge can theoretically result in 80% or higher probability of detecting noncompliant discharge concentrations of 12 vs. 10 live organisms m^{-3}. Thus, pooling volumes from separate trials will allow lower concentrations to be differentiated from the performance standard, although the practicability and economic costs of doing so have not been evaluated. Moreover, the practical limits of increased statistical sample sizes may already tax the capabilities of well-engineered land-based ballast water test facilities used in verification testing. Shipboard testing in the U.S. has been done on a pilot scale to date (i.e., the USCG Shipboard Testing Evaluation Program, STEP), but we imagine that pooling volumes from multiple trials might also be problematic on vessels used for shipboard verification testing and compliance testing. According to Table C-1, to meet a standard ten-fold more stringent than the IMO D-2/ Phase 1 standard would require anywhere from 120-600 m^3 of whole-water sample volumes, which is impracticable at this point – test facilities in the U.S. typically analyze ~5 m^3 of water per test (e.g., MERC 2009a, 2010a, 2010b; Great Ships Initiative 2010).

Additional Challenges of Sampling Large Volumes

Lee et al. (2010) calculated the probability of finding one or more organisms in a sample as $1\text{-}e^{c*v}$ (1 minus the probability of finding no organisms) for a series of organism concentrations and sample volumes, where *e* is the natural log, *c* is the true concentration of organisms, and *v* is the sample volume (Table C-3). The authors used the following assumptions:

- Performance standards are for the concentration of organisms in the ballast discharge (rather than the maximum number of organisms), so that the purpose of sampling is to estimate the "true" concentration of organisms in the discharge, referred to as average-based sampling;
- The organisms are randomly distributed and therefore amenable to modeling with the Poisson distribution, as above;
- All organisms are counted, with no human or instrumentation errors, so that any variation among samples for a given population (species) is from the natural stochasticity of sampling;
- The sample volume is calculated from the total volume of ballast water filtered (concentrated) and the filtrate volume that is subsampled. For example, following Lemieux et al. (2008): 100 m^3 of ballast water is filtered through a net to retain the zooplankton-sized organisms; the organisms are rinsed from the net, collected, and diluted to 1 L of water to give a concentration factor of 100,000:1. The organisms from 20 1-mL subsamples are counted: Total sample volume = 20 mL subsamples/1000 mL concentrated sample x 100 m^3 ballast water filtered = 2 m^3.

As Table C-2 illustrates, 100 L of ballast must be sampled to have a > 99% probability of detecting at least 1 zooplankton-sized organism when the true concentration is 100 organisms per m^3. When small sample volumes are collected, the probability of detecting an organism is low even at relatively high organism concentrations; for example, organisms will be detected in fewer than 10% of subsamples if a 1-L sample is taken and the "true" concentration is 100 organisms m^{-3}. This analysis also illustrates that when no organisms are detected from a relatively small sample, the true concentration in the ballast tank may still actually be large – it depends on the sample volume collected.

Lee et al. (2010) then estimated the upper possible concentration (UPC, upper 95% CI) of organisms actually present in ballast water from the number of zooplankton-sized organisms in a sample volume (ranging from 100 mL to 100 m^3) based on the Poisson distribution. As Table C-3 shows, 0 organisms detected in 1 m^3 of sample could correspond to a true concentration of organisms in the ballast tank of up to ~3.7 organisms m^{-3}. The error is much larger for a small sample volume of 1 L; 0 organisms detected could correspond to a true concentration of ~3,700 organisms m^{-3}.

Third, in the above analyses, the true concentrations of zooplankton-sized organisms are known. The goal in sampling unknown concentrations of organisms in ballast water is to accurately assess whether a given BWMS treats water with true organism concentrations that meet a given performance standard. Inherent stochasticity of sampling may result in an indeterminate category, as well, and the probability of obtaining an indeterminate evaluation increases with decreasing sample volume and increasing stringency of the ballast water standard (Figure C-1). Based on this analysis, it would be necessary to sample ~0.4 m^3 of ballast water to determine whether the IMO standard of < 10 zooplankton-sized organisms m^{-3} was met if fewer than approximately 10 organisms were observed in the sample (Figure C-1B).

Table C-2. Probability of detecting ≥ 1 zooplankton-sized organism for sample volumes (100 mL to 300 m^3) and ballast water concentrations (0 to 100 organisms m^{-3}). Gray boxes indicate probabilities of detection ≥ 0.95

	True concentration (organisms per m^3)						
Sample volume, m^3	**0**	**0.001**	**0.01**	**0.1**	**1**	**10**	**100**
0.0001 (100 mL)	0	<0.001	<0.001	<0.001	<0.001	0.001	0.01
0.001 (1 L)	0	<0.001	<0.001	<0.001	0.001	0.01	0.095
0.01 (10 L)	0	<0.001	<0.001	0.001	0.01	0.095	0.632
0.1 (100 L).	0	<0.001	0.001	0.01	0.095	0.632	>0.99
1	0	0.001	0.01	0.095	0.632	>0.99	>0.99
5	0	0.005	0.049	0.393	>0.99	>0.99	>0.99
10	0	0.010	0.095	0.632	>0.99	>0.99	>0.99
25	0	0.025	0.221	0.918	>0.99	>0.99	>0.99
50	0	0.049	0.393	>0.99	>0.99	>0.99	>0.99
100	0	0.095	0.632	>0.99	>0.99	>0.99	>0.99
300	0	0.259	0.950	>0.99	>0.99	>0.99	>0.99

Source: Lee et al. 2010.

Table C-3. Upper possible concentration (UPC) of zooplankton- sized organisms based on one and two tailed 95% exact confidence intervals when zero organisms are detected in a range of sample volumes

	Upper possible concentration, org m^{-3}	
Sample volume, m^3	**one-tailed**	**two-tailed**
0.0001 m^3 (100 mL)	29,960	36,890
0.001 m^3 (1 L)	2,996	3,689
0.01 m^3 (10 L)	299.6	368.9
0.1 m^3 (100 L)	29.96	36.89
0.5 m^3 (500 L)	5.992	7.378
1 m^3	2.996	3.689
10 m^3	0.300	0.369
100 m^3	0.030	0.037

Source: Lee et al. 2010.

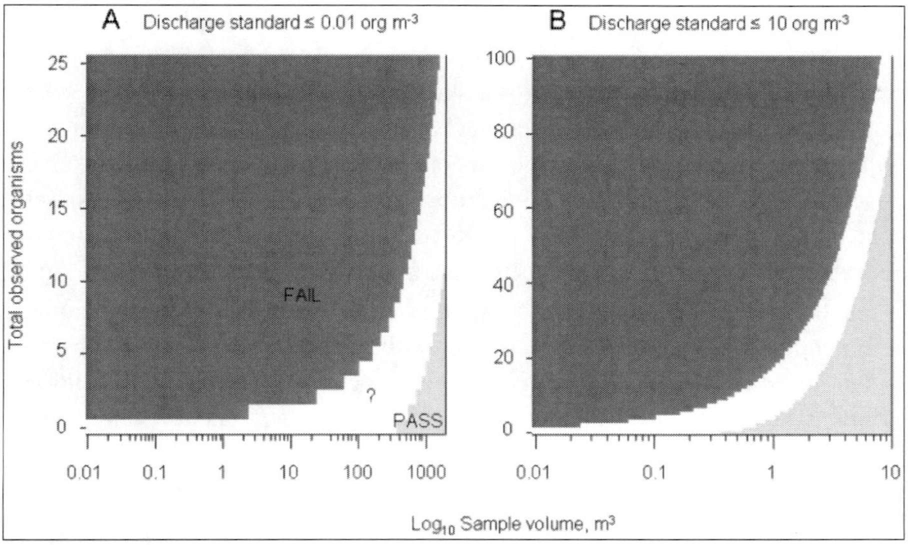

Source: Lee et al. 2010.

Figure C-1. Determining whether ballast water discharge exceeds or meets a performance standard of < 0.01 (A) and <10 (B) organisms m-3 (note: axes have different scales). Red regions indicate total organism counts that exceed the standard. Green regions indicate total organism counts that meet the standard. White regions indicate indeterminate results; counts in this region do not pass or fail inspection based on two-tailed 95% confidence intervals.

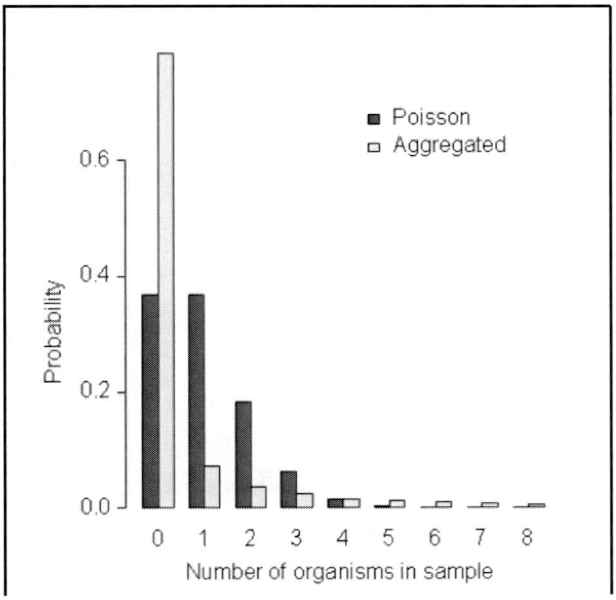

Source: Lee et al. 2010.

Figure C-2. Comparison of sample probabilities from a randomly distributed population (Poisson distribution) vs. an aggregated population with a dispersion parameter of 0.1 (negative binomial distribution) for a sample volume of 1 m3 and concentration of 1 organism m-3. For low organism numbers (3 or fewer m-3), the probability that a sample will contain zero organisms tends to be much greater for the aggregated population.

Spatially Aggregated Populations – Negative Binomial Distributions

This section illustrates how difficult statistical analyses can become when working with spatially aggregated populations. It further emphasizes the gains made from doing a complete count of a representative sample that has been continuously and isokinetically taken.

If organisms are aggregated (i.e., in clumped or contagious populations) rather than randomly distributed in a ballast tank, a different statistical approach is required. For aggregated populations, the variance exceeds the mean (negative binomial distribution, $\sigma^2 > \mu$); thus, as the variance increases, the number of organisms in a random sample is increasingly unpredictable. Because it is more difficult to accurately estimate the true concentration, more intensive sampling is required. Lee et al. (2010) recommend use of the negative binomial distribution to model aggregated populations. This distribution can be used to predict the probability of finding a certain number of organisms in a sample. It is defined by the mean (μ) and the dispersion or size parameter ($\theta = \mu^2/(\sigma^2 - \mu)$, where σ^2 = the variance; the smaller the dispersion parameter, the more aggregated the population.

The problem of having to sample multiple subsamples from large volumes to accurately assess low densities of organisms is compounded by aggregated distributions (Figure C-2). In the comparison given in Lee et al. (2010), for a randomly distributed population with a true concentration of 1 zooplankton- sized organism m^{-3}, ~37% of the subsamples from a 1 m^3 sample of treated ballast water would contain zero zooplankton-sized organisms. For an aggregated population with a dispersion parameter of 0.1, however, ~79% of the subsamples would contain zero organisms (Figure C-2). The relationship between the probability of finding zero organisms in a sample and the amount of aggregation is also illustrated (Figure C-3) for the concentration of 1 organism m^{-3}. As variance (σ^2) increases, the dispersion parameter θ decreases, indicating more aggregation, with increasing probability of finding no organisms in a sample. With more aggregation, the probability of samples containing large numbers of organisms relative to the true concentration also increases. Thus, large numbers of subsamples from large sample volumes must be taken to account for aggregated populations; otherwise, there will be a high probability that the concentration estimates from sample analyses will be either much lower or much higher than the true concentration.

Determination of whether a population is aggregated is complicated, since it is the scale of the aggregation pattern relative to the size of the sampling unit that controls the estimate of aggregation (Figure C-3). If organisms form clumps that are randomly distributed, the population may be highly aggregated, but in a small sample volume containing 0 or 1 organisms, the population will appear randomly distributed or only slightly aggregated. With increasing sample volume, the variance in the number of organisms increases in comparison to the mean, and maximum variance is encountered when the sample volume is equal to the volume of a single cluster of organisms (Elliott 1971). For larger sample volumes, a sample unit will include several clusters, so the variance decreases in comparison to the mean and the observations will approach a Poisson distribution. Lee et al. (2010) recommend the Taylor power law (Taylor 1961) as an alternative to the negative binomial, because it can accommodate a wider range of aggregated distributions than the negative binomial.

Overall, the possibility for and degree of aggregation represent challenges in sampling sufficiently large volumes of ballast water to determine whether a given BWMS passes or fails to meet standards more stringent than the present IMO guidelines, even if the true concentrations of organisms are 10 to 1,000 times higher than the performance standard. This remains a problem in quantifying many protist-sized organisms, but becomes less of a problem with very small organisms such as bacteria, which have a tendency to clump but are effectively counted as colonies and not individuals. However, in Lemieux et al. (2008), data from protist-sized microbeads at various concentrations were analyzed and concentrations of 100 mL^{-1} and lower were found to adhere to a Poisson distribution. The flasks of microbeads were well mixed, as would be samples of ballast water collected from the sample ports and collected to be representative of the entire volume sampled (e.g., over the entire discharge operation of the tank). Likewise, monocultures of protists in low densities (~10 to 30 mL^{-1}) adhered to a Poisson distribution (Nelson et al. 2009). These data lend support to using the Poisson distribution to analyzed ballast water samples.

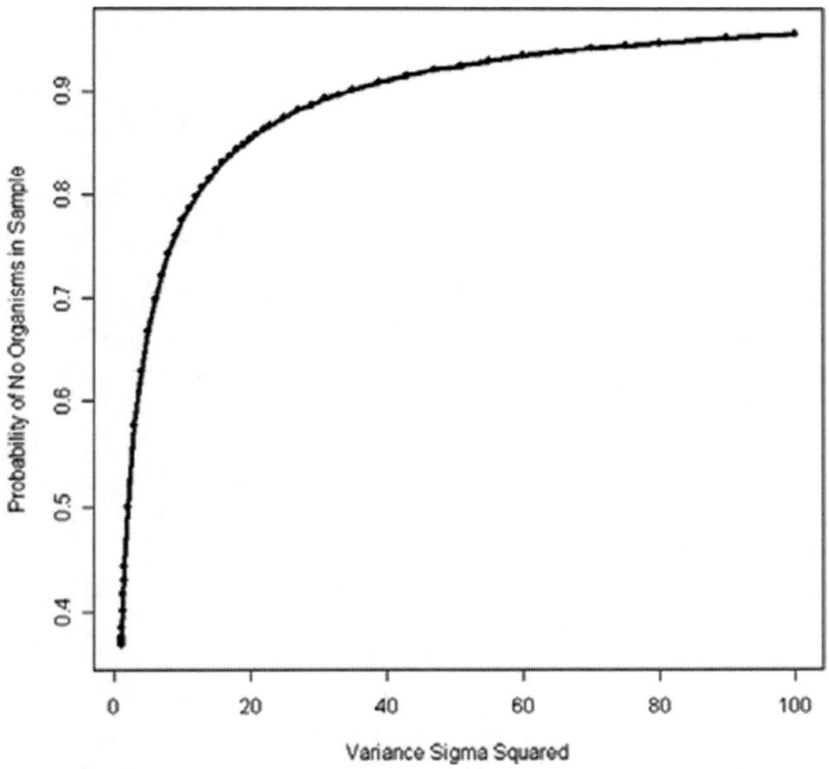

Figure C-3. The probability of finding zero organisms in a sample volume of 1 m^3 and concentration of $\mu = 1$ organism m^{-3}. The probability of 0 organisms = $(1 + \theta)^{-\theta}$, where dispersion parameter $\theta = 1/(\sigma^2 - 1)$. When $\sigma^2 = 1.0$, organisms are randomly distributed, at which the probability of 0 organisms in the sample = 0.37 (Poisson distribution) (Elliott 1971).

REFERENCES

Agusti, S. & Sánchez, M. C. (2002). Cell viability in natural phytoplankton communities quantified by a membrane permeability probe. *Limnology and Oceanography*, *47*, 818-828.

Albert, R. & Everett, R. (2010). *Capabilities of shipboard ballast water treatment technology*. Presentation to EPA's Science Advisory Board, July 29, 2010, Washington, DC.

Albert, R., Everett, R., Lishman, J. & Smith, D. (2010). *Availability and Efficacy of Ballast Water Treatment Technology: Background and Issue Paper*. U.S. EPA and US Coast Guard, Washington, DC. Available on the SAB website, at http://yosemite.epa.gov/sab/sabproduct.nsf/02ad90b136fc21ef85256eba00436459/9E6C799DF254393A8525762C004E60FF/$File/OW_Paper_Ballast_water_technology_issues_and_background_June_2010. pdf.

American Bureau of Shipping (ABS). (2010). *Ballast Water Treatment Advisory.* American Bureau of Shipping, Houston, TX.

American Public Health Association (APHA), American Water Works Association and Water Environment Federation. 2008. *Standard Methods for the Examination of Water and Wastewater*, 21st edition. APHA, Washington, DC.

Anderson, A. B., Johannessen, B. O. & Dragsund, E. (2004). Ballast water treatment verification protocol – DNV. In: *Proceedings of the 2nd International Ballast Water Treatment R&D Symposium*, by J.T. Matheickal and S. Raaymakers (eds.). IMO, London, UK.

Anderson, A., Cordell, J., Dobbs, F., Herwig, R. & Rogerson, A. (2008). *Screening of Surrogate Species for Ballast Water Treatment*. Final Report to NSF International, Ann Arbor, MI.

Aquatic Sciences. (1996). *Examination of Aquatic Nuisance Species Introductions to the Great Lakes through Commercial Shipping Ballast Water and Assessment of Control Options Phase II.* Final Report. A Report for the Canadian Coast Guard. Aquatic Sciences Inc., St. Catharines, Ontario, Canada, 60.

Audemard, C., Rease, K. C. & Burreson, E. M. (2004). Real-time PCR for detection and quantification of the protistan parasite *Perkinsus marinus* in environmental waters. *Applied and Environmental Microbiology*, *70*, 6611-6618.

Australian Quarantine and Inspection Service (AQIS). (1992). *Controls on the Discharge of Ballast Water and Sediment from Ships Entering Australia from Overseas*. AQIS Notice (Barrier Co-ordination) 92/2. AQIS, Canberra, Australia.

Australian Quarantine and Inspection Service (AQIS). (1993a). *Ballast Water Treatment for the Removal of Marine Organisms*. Prepared by Gutheridge, Haskins and Davey Pty Ltd. Ballast Water Research Series Report No. 1, AQIS, Canberra, Australia, 106.

Australian Quarantine and Inspection Service (AQIS). (1993b). *Ballast Water Management Study*. Prepared by Thompson Clarke Shipping Pty Ltd. Ballast Water Research Series Report No. 4, AQIS, Canberra, Australia, 254.

Avnimelech, Y., Troeger, B. W. & Reed, L. W. (1982). Mutual flocculation of algae and clay evidence and implications. *Science*, *216*, 63-65.

Baldino, R. (1995). *Outline of Standard Protocols for Particulate Organic Carbon (POC) Analyses*. University of Wisconsin - Madison, Madison, WI.

Barcina, I., Lebaron, P. & Vives-Rego, J. (1997). Survival of allochthonous bacteria in aquatic systems: a biological approach. *FEMS Microbiology Ecology*, *23*, 1-9.

Behrens, Y. S. & Hunt, C. (2000). The arrival and spread of the European green crab, *Carcinus maenas*, in the Pacific Northwest. Dreissena: The Digest of National Aquatic Nuisance Species Clearinghouse, *11*, 1–7.

Binet, M. T. & Stauber, J. L. (2006). Rapid flow cytometric method for the assessment of toxic dinoflagellate cyst viability. *Marine Environmental Research*, *62*, 247-260.

Blenkinsopp, S. A. & Lock, M. A. (1990). The measurement of electron transport system activity in river biofilms. *Water Research*, *24*, 441-445.

Bolch, C. J. & Hallegraeff, G. M. (1993). Chemical and physical treatment options to kill toxic dinoflagellate cysts in ships' ballast water. *Journal of Marine and Environmental Engineering*, *1*, 23-29.

Bosch, A., Abad, F. X. & Pintó, R. M. (2005). Human pathogenic viruses in the marine environment, pp.109-131. In: *Oceans and Health: Pathogens in the Marine Environment*, by S. Belkin and R.R. Colwell (eds.). Springer, New York, NY.

Bott, N. J., K. M. Ophel-Keller, M. T. Sierp, Herdina, K. P. Rowling, A. C. McKay, M. G. K. Loo, J. E. Tanner, and M. R. Deveney. (2010). Toward routine, DNA-based detection methods for marine pests. *Biotechnology Advances*, *28*, 706-714.

Bowles, N. D., Pearl, H. W. & Tucker, J. (1985). Effective solvents and extraction periods employed in phytoplankton carotenoid and chlorophyll determination. *Canadian Journal of Fisheries and Aquatic Science*, *42*, 1127-1131.

Boyer, G. L. (2007). The occurrence of cyanobacterial toxins in New York lakes: Lessons from the MERHAB-Lower Great Lakes Program. *Lake and Reservoir Management*, *23*, 153-60

Boyer, J. N., Kelbe, C. R., Ortner, P. B. & Rudnick, D. T. (2009). Phytoplankton bloom status: Chlorophyll *a* biomass as an indicator of water quality condition in the southern estuaries of Florida, USA. *Ecological Indicators*, *9*s, s56-s67.

Bratbak, G., Haslund, O. H., Heldal, M., Naess, A. & Roeggen, T. (1992). Giant marine viruses? *Marine Ecology Progress Series*, *85*, 201-202.

Brown & Caldwell, and Bay Engineering. Inc. (2008). *Port of Milwaukee Onshore Ballast Water Treatment – Feasibility Study Report. Phase 2*. Prepared for the Wisconsin Department of Natural Resources. Brown & Caldwell, Milwaukee, WI, 98.

Brown & Caldwell. (2007). *Port of Milwaukee Onshore Ballast Water Treatment – Feasibility Study Report.* Prepared for the Wisconsin Department of Natural Resources. Brown & Caldwell, Milwaukee, WI, 114.

Brussaard, C. P. D., Marie, D., Thyrhaug, R. & Bratbak, G. (2001). Flow cytometric analysis of phytoplankton viability following viral infection. *Aquatic Microbial Ecology*, *26*, 157-166.

Buchanan, C., Lacouture, R., Marshall, H. G., Olson, M. & Johnson, J. (2005). Phytoplankton reference communities for Chesapeake Bay and its tidal tributaries. *Estuaries*, *28*, 138-159.

Burkholder, J. M. & Gilbert, P. M. (2006). Intraspecific variability: An important consideration in forming generalizations about toxigenic algal species. *South African Journal of Marine Science*, *28*, 177-180.

Burkholder, J. M. (1992). Phytoplankton and episodic suspended sediment loading: Phosphate partitioning and mechanisms for survival. *Limnology and Oceanography*, *37*, 974-988.

Burkholder, J. M. (1998). Implications of harmful marine microalgae and heterotrophic dinoflagellates in management of sustainable marine fisheries. *Ecological Applications*, *8*, S37-S62.

Burkholder, J. M. (2009). Harmful algal blooms, pp. 264-285. In: *Encyclopedia of Inland Waters*, Volume *1*, by G.E. Likens (ed.). Elsevier, Oxford, UK.

Burkholder, J. M., Gordon, A. S., Moeller, P. D., Law, J. M., Coyne, K. J., Lewitus, A. J., Ramsdell, J. S., Marshall, H. G., Deamer, N. J., Cary, S. C., Kempton, J. W., Morton, S. L. & Rublee, P. A. (2005). Demonstration of toxicity to fish and to mammalian cells by *Pfiesteria* species: Comparison of assay methods and multiple strains. *Proceedings of the National Academy of Sciences (U.S.A)*, *102*, 3471-3476.

Burkholder, J. M., Hallegraeff, G. M., Melia, G., Cohen, A., Bowers, H.A., Oldach, D. W., Parrow, M. W., Sullivan, M. J., Zimba, P. V., Allen, E. H. & Mallin, M. A. (2007). Phytoplankton and bacterial assemblages in ballast water of U.S. military ships as a function of port of origin, voyage time and ocean exchange practices. *Harmful Algae*, *6*, 486-518.

California Association of Port Authorities (CAPA). (2000). F*easibility of Onshore Ballast Water Treatment at California Ports*. Prepared by URS Corporation/Dames & Moore. CAPA, Sacramento, CA.

California State Lands Commission (SLC). (2009). *Assessment of the Efficacy, Availability and Environmental Impacts of Ballast Water Treatment Systems for Use in California Waters*. California SLC, Sacramento, CA, 180.

California State Lands Commission (SLC). (2010). *2010 Assessment of the Efficacy, Availability and Environmental Impacts of Ballast Water Treatment Systems for Use in California Waters*. CSLC, Sacramento, CA, 141.

California State Water Resources Control Board (SWRCB). (2002). *Evaluation of Ballast Water Treatment Technology for Control of Nonindigenous Aquatic Organisms*. SWRCB, California Environmental Protection Agency, Sacramento, CA, 70 pp.

Canadian Marine Advisory Council (CMAC). (2000). *Guidelines for the Control of Ballast Water Discharge from Ships in Waters under Canadian Jurisdiction*. TP 13617, CMAC, Ottawa, Canada.

Canadian Marine Advisory Council (CMAC). (2007). *A Guide to Canada's Ballast Water Control and Management Regulations*. TP 13617E, Transports Canada, Ottawa, Canada.

Cangelosi, A. A., Mays, N. L., Balcer, M. D., Reavie, E. D., Reid, D. M., Sturtevant, R. & Gao, X. (2007). The response of zooplankton and phytoplankton from the North American Great Lakes to filtration. *Harmful Algae*, *6*, 547-566.

Carlton, J. T., Reid, D. M. & van Leeuwen, H. (1995). *Shipping Study. The Role of Shipping in the Introduction of Nonindigenous Aquatic Organisms to the Coastal Waters of the United States (other than the Great Lakes) and an Analysis of Control Options*. Report No. CG-D-11-95, U. S. Coast Guard, Groton CT and U. S. Dept. of Transportation, Washington DC.

Caron, D. A., Countway, P. D. & Brown, M. V. (2004). The growing contributions of molecular biology and immunology to protistan ecology: Molecular signatures as ecological tools. *Journal of Eukaryotic Microbiology*, *51*, 38-48.

Cohen, A. N. & Foster, B. (2000). The regulation of biological pollution: Preventing exotic species invasions from ballast water discharged into California coastal waters. *Golden Gate University Law Review*, *30*, 787-883.

Cohen, A. N. (1998). *Ships' Ballast Water and the Introduction of Exotic Organisms into the San Francisco Estuary: Current Status of the Problem and Options for Management.* A report for the Collaboration Among State and Federal Agencies to Improve California's

Water Supply (CALFED) and the California Urban Water Agencies. San Francisco Estuary Institute, Richmond, CA.

Colnar, A. M. & Landis, W.G. (2007). Conceptual model development for invasive species and a regional risk assessment case study: the European Green Crab, *Carcinus maenas*, at Cherry Point, Washington USA. *Human and Ecological Risk Assessment*, *13*, 120-155.

Connelly, S. J., Wolyniak, E. A., Dieter, K. L., Williamson, C. E. & Jellison, K.L. (2007). Impact of zooplankton grazing on the excystation, viability, and infectivity of the protozoan pathogens *Cryptosporidium parvum* and *Giardia lamblia. Applied and Environmental Microbiology*, *73*, 7277-7282.

Cromeans, T., Narayanan, J., Jung, K., Ko, G., Wait, D. & Sobsey, M. (2005). *Development of Molecular Methods to Detect Infections Viruses in Water.* American Water Works Association Research Foundation Report 90995F. ISBN 9781843398912, 128.

Cuker, B. E. & Hudson Jr., L. (1992). Type of suspended clay influences zooplankton response to phosphorus loading. *Limnology and Oceanography*, *37*, 566-576.

Dames & Moore. (1998). *Ballast water management technical memorandum for the Port of Oakland.* Job No. 02801-028-086, November 6, 1998. Dames & Moore, San Francisco, CA.

Dames & Moore. (1999). *Phase I Final Report Ballast Water Exchange and Treatment.* A Report for the California Association of Port Authorities, Pacific Merchant Shipping Association, Steamship Association of Southern California, and Western States Petroleum Association. Dames & Moore, San Francisco, CA.

Darzynkiewicz, Z., Li, X. & Gong, J. (1994). Assays of cell viability: Discrimination of cells dying by apoptosis. In: *Methods in Cell Biology*, by Z. Darzynkiewicz, J.P. Robinson, and H.A. Crissman (eds.). Academic Press, New York.

Deines, A. M., Chen, V. & Landis, W. G. (2005). Modeling the risks of non-indigenous species introductions using a patch-dynamics approach incorporating contaminant effects as a disturbance. *Risk Analysis*, *6*, 1637-1651.

Dobbs, F. C. & Rogerson, A. (2005). Ridding ships' ballast water of microorganisms. *Environmental Science and Technology*, *39*, 259A-264A.

Doblin, M. A. & Dobbs, F. C. (2006). Setting a size-exclusion limit to remove toxic dinoflagellate cysts in ships' ballast water. *Marine Pollution Bulletin*, *52*, 259-263.

Dobroski, N., Scianni, C., Gehringer, D. & Falkner, M. (2009). *2009 Assessment of the Efficacy, Availability and Environmental Impacts of Ballast Water Treatment Systems for Use in California Waters.* Produced for the California State Legislature. Available at: http://www.slc.ca.gov/spec_pub/mfd/ballast_water/documents/2009cslctechreportfinal.pdf.

Dobroski, N., Scianni, C., Takata, L., & Falkner, M. (2009). October 2009 update: Ballast water treatment technologies for use in California waters. California State Lands Commission, Marine Invasive Species Program. Produced for the California State Legislature. Final Report available at: http://www.slc.ca.gov/Spec_Pub/MFD/Ballast_Water/Reports_Presentations.html.

Donner, P. (2010a). Ballast Water Treatment Ashore Brings More Benefits. Pages 97-105 in: Emerging Ballast Water Management Systems. Proceedings of the IMO-WMU Research and Development Forum, 26-29 January 2010, Malmö, Sweden. A GloBallast-Global Industry Alliance and World Maritime University Initiative. Bellefontaine, N., F. Haag, O. Lindén and J. Matheickal (eds.). WMU Publication, Malmö, Sweden.

Donner, P. (2010b). Ballast water treatment ashore—better for the environment and for seafarers. WMU Journal of Maritime Affairs, *9*(2), 191-199.

Donner, P. (2010c). Is there a case for shore-based ballast water treatment facilities? 5th International Conference on Ballast Water Management 2010. Singapore, Nov. 1-4, 2010. Session 1: Ballast Water Management Plans by Ports and Flag States.

Drake, J. M. & Lodge, D.M. (2006). Allee effects, propagule pressure and the probability of establishment: risk analysis for biological invasions. *Biological Invasions*, *8*, 365–375.

Drake, J. M. (2004). Allee Effects and the Risk of Biological Invasion. *Risk Analysis*, *24*, 795-802.

Drake, J. M. (2005). Risk Analysis For Invasive Species And Emerging Infectious Diseases: Concepts And Applications. *Am. Midl. Nat.*, *153*, 4–19.

Drake, J. M., Lodge, D. M. & Lewis, M. (2005). Theory and preliminary analysis of species invasions from ballast water: controlling discharge volume and location. *American Midland Naturalist*, *154*, 459-470.

Elliot, J. M. (1971). *Some Methods for the Statistical Analysis of Samples of Benthic Invertebrates.* Scientific Publication No. 25, Freshwater Biological Association, Windermere, UK.

Fout, G. S., Schaefer III, F. W., Messer, J. W., Dahling, D. R. & Stetler, R. E. (1996). *ICR Microbial Laboratory Manual.* U.S. Environmental Protection Agency, Washington, DC.

Garvey, M., Moriceau, B. & Passow, U. (2007). Applicability of the FDA assay to determine the viability of marine phytoplankton under different environmental conditions. *Marine Ecology Progress Series*, *352*, 17-26.

Gauthier, D. & Steel, D. A. (1996). *A Synopsis of the Situation Regarding the Introduction of Nonindigenous Species by Ship-Transporting Ballast Water in Canada and Selected Countries.* Canadian Manuscript Report of Fisheries and Aquatic Sciences No. 2380. Marine Environmental Sciences Division, Department of Fisheries and Oceans, Maurice Lamontagne Institute, Mont-Joli (Quebec), Canada.

Glosten Associates. (2002). *Pacific Ballast Water Treatment Pilot Project: Ballast Water Transfer Study: Technical Feasibility with Associated Capital Costs*. Report prepared for the Port of Seattle, Seattle, WA. File No. 01080. Glosten Associates, Seattle, WA, 36.

Great Ships Initiative. (2010). Report of the Land-Based Freshwater Testing by the Great Ships Initiative of the Siemens SiCURETM Ballast Water Management System for Type Approval According to Regulation D-2 and the Relevant IMO Guidelines. Great Ships Initiative. http://www.nemw. org/GSI/GSI-LB-F-A-1.pdf.

Greenman, D., Mullen, K. & Parmar, S. (1997). *Ballast Water Treatment Systems: A Feasibility Study*. Produced in cooperation with ENS Chris Freise, U.S. Coast Guard Office of Response. Commissioned by the U.S. Coast Guard and Worcester Polytechnic Institute. Submitted to Professor Matthew Ward, Project Center, Worcester Polytechnic Institute, Washington, DC, 59.

Gregg, M. D. & Hallegraeff, G. M. (2007). Efficacy of three commercially available ballast water biocides against vegetative microalgae, dinoflagellate cysts and bacteria. *Harmful Algae*, *6*, 567-584.

Griffin, D. W., Donaldson, K. A., Paul, J. H. & Rose, J. B. (2003). Pathogenic human viruses in coastal waters. *Clinical Microbiology Reviews*, *16*, 129–143.

Guy, R. A., Payment, P., Krull, U. J. & Horgen, P. A. (2003). Real-time PCR for quantification of *Giardia* and *Cryptosporidium* in environmental water samples and sewage. *Applied and Environmental Microbiology*, *69*, 5178-5185.

Hallegraeff, G. M. & Bolch, C. J. (1991). Transport of toxic dinoflagellate cysts via ships' ballast water. *Marine Pollution Bulletin*, *22*, 27-30.

Hallegraeff, G. M. & Bolch, C. J. (1992). Transport of dinoflagellate cysts in ship's ballast water: Implications for plankton biogeography and aquaculture. *Journal of Plankton Research*, *14*, 1067-1084.

Hallegraeff, G. M., Valentine, J. P., Marshall, J. -A. & Bolch, C. J. (1997). Temperature tolerances of toxic dinoflagellate cysts: application to the treatment of ships' ballast water. *Aquatic Ecology*, *31*, 47-52.

Hampel, M., Moreno-Garrido, I., Sobrino, C., Lubia, H. L. M. & Blasco, J. (2001). Acute toxicity of LAS homologues in marine microalgae: Esterase activity and inhibition growth as endpoints of toxicity. *Ecotoxicology and Environmental Safety*, *48*, 287-292.

Havelaar, A. H. (1994). Application of HACCP to Drinking Water Supply. *Food Control*, *5*(3), 145-152.

Hendry, G. A. F., Houghton, J. D. & Brown, S. B. (1987). The degradation of chlorophyll - a biological enigma. *New Phytologist*, *107*, 255-302.

Hilliard R. W. & Matheickal, J. T. (2010). Alternative Ballast Water Management Options for Caspian Region Shipping: Outcomes of a Recent CEP/IMO/UNOPS Project. Pages 119-138 in: Emerging Ballast Water Management Systems. Proceedings of the IMO-WMU Research and Development Forum, 26-29 January 2010, Malmö, Sweden. A GloBallast-Global Industry Alliance and World Maritime University Initiative. Bellefontaine, N., F. Haag, O. Lindén and J. Matheickal (eds.). WMU Publication, Malmö, Sweden.

Hilliard, R. W. (2006). Assessment of Shipping Traffic and Ballast Water Movements to and from Caspian Sea, and Preliminary Appraisal of Possible Ballast Water Management Options. Draft Report. Prepared for the International Maritime Organization. URS Corporation. Accessed at: http://www.caspianenvironment.org/Newsite/Activities-Ballast%20 Waters.htm

Hoff, R. M., Strachan, W. M. J., Sweet, C. W. Chan, C. H., Shackleton, M., Bidleman, T. F., Brice, K. A., Burniston, D. A., Cussion, S., Gatz, D. F., Harlin, K. & Schroeder, W. H. (1996). Atmospheric deposition of toxic chemicals to the Great Lakes: A review of data through 1994. *Atmospheric Environment*, *30*, 3505-3527.

Holm-Hansen, O. (1973). The use of ATP determinations in ecological studies. *Bulletin of the Ecological Research Commission (Stockholm)*, *17*, 215-222.

Howard, G., Bartram, J., Pedley, S., Schmoll, O., Chorus, I. & Berger, P. (2006). Groundwater and public health, pp. 3-19. In: *Protecting Groundwater for Health: Managing the Quality of Drinking-Water Sources*, by O. Schmoll, G. Howard, J. Chilton, and I. Chorus (eds.). IWA Publishing, London, U.K.

Hunt, C. D., Tanis, D. C., Stevens, T. G., Frederick, R. M. & Everett, R. A. (2005) (August 1). Verifying ballast water treatment performance. *Environmental Science and Technology*, 321A-328A.

International Maritime Organization (IMO). (1991). *International Guidelines for Preventing the Introduction of Unwanted Aquatic Organisms and Pathogens from Ships' Ballast Water and Sediment Discharges*. IMO, Marine Environment Protection Committee, Resolution MEPC, *50*(31), adopted 4 July 1991.

International Maritime Organization (IMO). (1993). Guidelines for Preventing the Introduction of Unwanted Aquatic Organisms and Pathogens from Ships' Ballast Water and Sediment Discharges. IMO, Resolution A. *774*(18), adopted 4 November 1993.

International Maritime Organization (IMO). (1997). *Guidelines for the Control and Management of Ships' Ballast Water to Minimize the Transfer of Harmful Aquatic Organisms and Pathogens*. IMO, 20th Assembly, Resolution A. *868*(20), adopted 27 November 1997.

International Maritime Organization (IMO). (2004). *International Convention for the Control and Management of Ships' Ballast Water and Sediments, 2004*. IMO, International Conference on Ballast Water Management for Ships.

International Maritime Organization (IMO). (2005a). *Guidelines tor Ballast Water Management Equivalent Compliance (G3)*. IMO, Marine Environment Protection Committee, Resolution MEPC, *123*(53), adopted 22 July 2005.

International Maritime Organization (IMO). (2005b). *Guidelines for Ballast Water Management and Development of Ballast Water Management Plans (G4)*. IMO, Marine Environment Protection Committee (MEPC), Resolution MEPC, *127*(53), adopted 22 July 2005.

International Maritime Organization (IMO). (2006). *Guidelines for Ballast Water Reception Facilities (G5)*. IMO, Marine Environment Protection Committee, Marine Environment Protection Committee (MEPC), Resolution MEPC, *153*(55), adopted 13 October 2006.

International Maritime Organization (IMO). (2008a). *Guidelines for Approval of Ballast Water Management Systems (G8)*. Marine Environment Protection Committee (MEPC), Annex 4 Resolution MEPC, *174*(58). Available at: http://www.regulations.gov/search/Regs/home.html# documentDetail?R=09000064807e8904.

International Maritime Organization (IMO). (2008b). *Procedure for Approval of Ballast Water Management Systems that Make Use of Active Substances (G9)*. Marine Environment Protection Committee (MEPC), Annex 1 Resolution MEPC.169(57). Available at: http://www.regulations.gov/search/Regs/home.html#documentDetail?R=09000064807e890e.

Inubushi, K., Brookes, P. C. & Jenkinson, D. S. (1989). Adenosine 5'-triphosphate and adenylates energy charge in waterlogged soil. *Soil Biology and Biochemistry*, *21*, 733-739.

Jeffrey, S. W. & Vest, M. (1997). Introduction to marine phytoplankton and their pigment signatures, 37-87. In: *Phytoplankton Pigments in Oceanography*, by S.W. Jeffrey, R.F.C. Mantoura, and S.W. Wright (eds.). The United Nations Educational, Scientific, and Cultural Organization (UNESCO), Paris, France.

Jeffrey, S., Mantoura, R. & Wright, S. (eds.). (1997). *Phytoplankton Pigments in Oceanography: Guidelines to Modern Methods*. UN Educational, Scientific & Cultural Organization, Paris, France.

Johnson, W. S. & Allen, D. M. (2005). *Zooplankton of the Atlantic and Gulf Coasts – A Guide to Their Identification and Ecology*. John Hopkins University Press, Baltimore, MD.

Jones, W. J., Preston, C. M., Marin III, R., Scholin, C. A. & Vrijenhoek, R. C. (2008). A robotic molecular method for in situ detection of marine invertebrate larvae. *Molecular Ecology Resources*, *8*, 540-550.

Kaneshiro, E. S., Wyder, M. A., Wu, Y. -P. & Cushion, M. T. (1993). Reliability of calceinacetoxy methyl ester and ethidiumhomodimer or propidium iodide for viability assessment of microbes. *Journal of Microbiological Methods*, *17*, 1-16.

Karl, D. M. (1980). Cellular nucleotide measurements and applications in microbial ecology. *Microbiological Reviews*, *44*, 739-796.

Karlson, B., Cusack, C. & Bresnan, E. (2010). *Microscopic and Molecular Methods for Quantitative Phytoplankton Analysis.* Intergovernmental Oceanographic Commission (IOC) Manuals and Guides #55. IOC, United Nations Educational, Scientific and Cultural Organization, Paris, France, 114 pp. Kay, S.H. 1985. Cadmium in aquatic food webs. *Residue Reviews*, *96*, 13-43.

Keller, R. P., Drake, J. M., Drew, M. B. & Lodge, D.M. (2010). Linking environmental conditions and ship movements to estimate invasive species transport across the global shipping network. *Diversity and Distributions* DOI: 10.1111/j.1472-4642/2010/00696x

Kieft, T. L. & Rosaker, L. L. (1991). Application of respiration and adenylates-based soil microbiological assays to deep subsurface terrestrial sediments. *Soil Biology and Biochemistry*, *23*, 563-568.

King, D. M. & Tamburri, M. N. (2010). Verifying compliance with ballast water discharge regulations. *Ocean Development & International Lawi, 41*, 152-165.

Kokinos, J. K., Eglington, T. I., Goni, M. A., Boon, J. J., Martoglio, P. A. & Anderson, D. M. (1998). Characterization of a highly resistant biomacromolecular material in the cell wall of a marine dinoflagellate resting cyst. *Organic Geochemistry*, *28*, 265-288.

Kolar, C. S. & Lodge, D. M. (2001). Progress in invasion biology: predicting invaders. *Trends Ecol Evol*, *16*, 199–204.

Kolar, C. S. & Lodge, D. M. (2002). Ecological predictions and risk assessment for alien species. *Science*, *298*, 1233–6.

Kolar, C. S. (2004). Risk assessment and screening for potentially invasive species. *New Zeal J Mar Fresh*, *38*, 391–7.

Kotinis, M., Parsons, M. G., Lamb, T. & Sirviente, A. (2004). Development and investigation of the ballast-free ship concept. *Transactions Society of Naval Architects and Marine Engineers*, *112*.

Kudela, R. M., Howard, M. D. A., Jenkins, B. D., Miller, P. E. & Smith, G. J. (2010). Using the molecular toolbox to compare harmful algal blooms in upwelling systems. *Progress in Oceanography*, *85*, 108-121.

Laabir, M. & Gentien, P. (1999). Survival of toxic dinoflagellates after gut passage in the Pacific oyster *Crassostreagigas*Thunberg. *Journal of Shellfish Research*, *18*, 217-222.

Landis, W. G. (2004). Ecological risk assessment conceptual model formulation for nonindigenous species. *Risk Analysis*, *24*, 847-858.

LeChevallier, M. W. & Au, K. K. (2004). *Water Treatment and Pathogen Control; Process Efficiency in Achieving Safe Drinking Water.* World Health Organization, Geneva. IWA Pubishing, London.

Lee II, H., Reusser, D. A., Frazier, M. & Ruiz, G. (2010). *Density matters: Review of approaches to setting organism-based discharge standards.* U.S. EPA, Office of Research and Development, National Health and Environmental Effects Research Laboratory, Western Ecology Division, Washington, DC. EPA/600/R-10/031.

Lemieux, E. J., Drake, L. A., Riley, S. C., Schrack, E. C., Hyland, W. B., Grant, J. F. & Moser, C. S. (2010). *Design and Preliminary Use of a Commercial Filter Skid to Capture*

Organisms ≥ 50 μm in Minimum Dimension (Nominally Zooplankton) for Evaluating Ships' Ballast Water Management Systems at Land- Based Test Facilities. Report 6130/1029. Center for Corrosion Science and Engineering, Naval Research Laboratory, Washington, DC, 41.

Lemieux, E. J., Robbins, S., Burns, K., Ratcliff, S. & Herring, P. (2008). *Evaluation of Representative Sampling for Rare Populations using Microbeads*. Report No. CG-D-03-09, United States Coast Guard Research and Development Center, Groton, CT.

Lloyd's Register. (2010). *Ballast Water Treatment Technology: Current Status*. Lloyd's Register, London.

Loganathan, B. G. & Kannan, K. (1994). Global organochlorine contamination trends: an overview. *Ambio*, *23*, 187-191.

Lund, J. W. G., Kipling, C. & LeCren, E. D. (1958). The inverted microscope method for estimating algal numbers and the statistical basis of estimations by counting. *Hydrobiologia*, *11*, 143-170.

MacIsaac, E. A. & Stockner, J. G. (1993). Enumeration of phototrophic picoplankton by autofluorescence microscopy, 187-198. In: *Handbook of Methods in Aquatic Microbial Ecology*, by P.F. Kemp, B.F. Sherr, E.B. Sherr, and J.J. Cole (eds.). Lewis Publishers, CRC Press, Inc., Boca Raton, FL.

Mackey, M. D., Mackey, D. J., Higgins, H. W. & Wright, S. W. (1996). CHEMTAX – A program for estimating class abundances from chemical markers: Application to HPLC measurements of phytoplankton. *Marine Ecology Progress Series*, *144*, 265-283.

Madigan, M. T. & Martinko, J.M. (2006). *Biology of Microorganisms*, 11th ed. Pearson Prentice Hall, Upper Saddle River, NJ, G-14.

Maier, R. M., Pepper, I. L. & Gerba, C. P. (eds.). *Environmental Microbiology*. Elsevier, New York, NY.

MARAD. (Feb. 2011). U.S. Water Transport Statistical Snapshot. US DOT Maritime Administration. http://www.marad.dot.gov/library_landing_ page/data_and_statistics/ Data_and_Statistics.htm

Marine Environment Protection Committee (MEPC). (21 April, 2006a). *Harmful Aquatic Organisms in Ballast Water - Basic Approval of Active Substances Used by Pure Ballast Management System*. International Maritime Organization, 55th session, Agenda Item 2, MEPC 55/2/5, International Maritime Organization.

Marine Environment Protection Committee (MEPC). (28 Feb., 2006b). 54th Session, Agenda item 2 – *Harmful Aquatic Organisms in Ballast Water* – Report of the first meeting of the Group of Experts on the Scientific Aspects of Marine Environmental Protection Border-Wide Workgroup (GESAMP-BWWG). MEPC 54/2/12, International Maritime Organization.

Maritime Environmental Resource Center (MERC). (2009a). Land-Based Evaluations of the Maritime Solutions, Inc. Ballast Water Treatment System. MERC ER02-09, [UMCES] CBL 09-138.

Maritime Environmental Resource Center (MERC). (2009b). *M/V* Cape Washington *Ballast Water Treatment Test Facility Validation*. Chesapeake Biological Laboratory, University of Maryland Center for Environmental Science, Solomons, MD.

Maritime Environmental Resource Center (MERC). (2009c). *Test Plan for the Performance Evaluation of the* Severn Trent De Nora *BalPure*^{TM} *BP-2000 Ballast Water Management*

System. Chesapeake Biological Laboratory, University of Maryland Center for Environmental Science, Solomons, MD.

Maritime Environmental Resource Center (MERC). (2010a). Land-Based Evaluations of the Siemens Water Technologies SiCURE Ballast Water Management System. MERC ER02-10, [UMCES] CBL 10-038.

Maritime Environmental Resource Center (MERC). (2010b). Land-Based Evaluations of the Severn Trent De Nora BalPure BP-1000 Ballast Water Management System. MERC ER01-10, [UMCES] CBL 10-015.

Marret, F. & Zonneveld, K. A. F. (2003). Atlas of modern organic-walled dinoflagellate cyst distribution. *Review of Palaeobotany and Palynology*, *125*, 1-200.

Matsuoka, K. & Fukuyo, Y. (2000). *Technical Guide for Modern Dinoflagellate Cyst Study*. Available at: http://dinos.anesc.u-tokyo.ac.jp/ technical_guide/main.pdf.

Miller, A. W., Frazier, M., Smith, G. E., Perry, E. S., Ruiz, G. M. & Tamburri, M.N. (2011). Enumerating Sparse Organisms in Ships' Ballast Water: Why Counting to 10 is Difficult? *Environmental Science and Technology*, *45*, 3539–3546.

Minton, M. S., Verling, E., Miller, A. W. & Ruiz, G. M. (2005). Reducing propagule supply and coastal invasions via ships: effects of emerging strategies. *Frontiers in Ecology and the Environment*, *3*, 304-308.

Moeller, P. D. R., Morton, S. L., Mitchell, B. A., Sivertsen, S. K., Fairey, E. R., Mikulski, T. M., Glasgow, H. B., Deamer-Melia, N. J., Burkholder, J. M. & Ramsdell, J. S. (2001). Current progress in isolation and characterization of toxins isolated from *Pfiesteria* spp. *Environmental Health Perspectives*, *109*, 739-743.

Montresor, M., Nuzzo, L. & Mazzocchi, M. G. (2003). Viability of dinoflagellate cysts after the passage through the copepod gut. *Journal of Experimental Marine Biology and Ecology*, *287*, 209-221.

Mosher, J. J., Levison, B. S. & Johnston, C. G. (2003). A simplified dehydrogenase enzyme assay in contaminated sediment using 2-(p-iodophenyl)-3(p-nitrophenyl)-5-phenyl tetrazolium chloride. *Journal of Microbiological Methods*, *53*, 411-415.

Munn, C. B. (2006). Viruses as pathogens of marine organisms – from bacteria to whales. *Journal of the Marine Biological Association of the UK*, *86*, 453–467.

Murphy, K. R., Ritz, D. & Hewitt, C. L. (2002). Heterogeneous zooplankton distribution in a ship's ballast tanks. *Journal of Plankton Research*, *24*, 729-734.

Myers, G. J. (1979). *The Art of Software Testing*. John Wiley & Sons, New York, NY, 12-13,128.

National Environmental Coalition on Invasive Species (NECIS). (July 2008). New Ballast Tank Legislation. http://www.pollutionengineering. com/Articles/Feature_Article/BNP_GUID_9-5-2006_A_10000000000000 380477

National Environmental Services Center (NESC). (2000a). *Slow Sand Filtration. Tech Brief: A National Drinking Water Clearinghouse Fact Sheet*. National Environmental Services Center, West Virginia University. Morgantown, WV.

National Environmental Services Center (NESC). (2000b). *Ultraviolet Disinfection. Tech Brief: A National Drinking Water Clearinghouse Fact Sheet*. National Environmental Services Center, West Virginia University. Morgantown, WV.

National Research Council (NRC) Marine Board. (1996). Stemming the Tide: Controlling Introductions of Nonindigenous Species by Ships' Ballast Water. Committee on Ships'

Ballast Operations, Marine Board, Commission on Engineering and Technical Systems, NRC. National Academy Press, Washington, D.C.

National Science Foundation (NSF) (2003). *Engineering Controls for Ballast Water Discharge: Developing Research Needs*. Report of a workshop held on April 28-30, Seattle, WA. Bioengineering and Environmental Systems Division of the Engineering Directorate, NSF, Arlington, VA.

Nelson, B. N., Lemieux, E. J., Drake, L., Anderson, D., Kulis, D., Burns, K., Anderson, D., Welschmeyer, N., Smith, S., Scianni C, C., Weir, T., Riley, S. & Herring, P. (2009). *Phytoplankton Enumeration and Evaluation Experiments*. Report No. CG-D-06-09, United States Coast Guard Research & Development Center, Groton, CT.

New Zealand Ministry of Agriculture and Forestry (NZMAF). (1998). Import Health Standard For Ships' Ballast Water From All Countries (Biosecurity Act 1993). NZ MAF, Wellington, New Zealand.

New Zealand Ministry of Agriculture and Forestry (NZMAF). (2005). Import Health Standard For Ships' Ballast Water From All Countries. Issued pursuant to Section 22 of the Biosecurity Act 1993. NZ MAF, Wellington, New Zealand.

Notermans, S., Gallhoff, G., Zweitering, M. H. & Mead, G. C. (1994). The HACCP concept: Identification of potentially hazardous micro-organisms. *Food Microbiology* II, 203-214.

Oemke, D. (1999). *The Treatment of Ships' Ballast Water*. EcoPorts Monograph Series No. 18, Ports Corporation of Queensland, Brisbane, Australia, 102.

Ogilvie, D. J. (1995). Land-based treatment options. Pages 113-123 in: *Ballast Water. A Marine Cocktail on the Move. Proceedings of the National Symposium*, 27-29 June 1995, Wellington, New Zealand. The Royal Society of New Zealand. Miscellaneous Series 30.

Ogilvie, D. J. (1995). Land-based treatment options. Pages 113-123 in: Ballast Water. A Marine Cocktail on the Move. Proceedings of the National Symposium, 27-29 June 1995, Wellington, New Zealand. The Royal Society of New Zealand. Miscellaneous Series 30.

Oliver, J. D. (1993). Formation of viable but nonculturable cells, 239-272. In: *Starvation in Bacteria*, by S. Kjelleberg (ed.). Plenum Press, New York, NY.

Oudot C., Gerard, R., Morin, P. & Gningue, I. (1988). Precise shipboard determination of dissolved oxygen (Winkler procedure) for productivity studies with a commercial system. *Limnology and Oceanography*, *33*, 146-150.

Parshionikar, S., Laseke, I. & Fout, G. S. (2010). Use of propidiummonoazide in reverse transcriptase PCR to distinguish between infectious and noninfectious enteric viruses in water samples. *Applied and Environmental Microbiology*, *76*, 4318-4326.

Parsons, M. G. (1998). Flow-through ballast water exchange. *Society of Naval Architects and Marine Engineers (NAME) Transactions*, *106*, 485-493.

Parsons, M. G. (2010). The Variable Buoyancy Ship: A Road to the Elimination of Ballast. Proceedings from IMO and World Maritime University Research and Development Forum: Emerging Ballast Water Management Systems, January 26-29, Malmo, Sweden.

Pate, A. S., De Souza, A. E. & Farrow, D. R. G. (1992). *Agricultural Pesticide Use in Coastal Areas: A National Summary*. NOAA, Rockville, MD.

Peal, D. B. (1975). *Phthalate Esters: Occurrence and Biological Effects*. Springer-Verlag, New York, NY, 1-41.

Pereira, N. N., Botter, R. C., Brinati, H. L. & Trevis, E. F. (2010). A Study of Ballast Water Treatment Applied on Iron Ore Ports in Brazil Using Discrete Simulation. Pages 77-91 in: Emerging Ballast Water Management Systems. Proceedings of the IMO-WMU

Research and Development Forum, 26-29 January 2010, Malmö, Sweden. A GloBallast-Global Industry Alliance and World Maritime University Initiative. Bellefontaine, N., F. Haag, O. Lindén and J. Matheickal (eds.). WMU Publication, Malmö, Sweden.

Phillips, S. (2006). *Ballast Water Issue Paper*. Report for the Pacific States Marine Fisheries Commission. Portland, OR. 34 pp. (http://www.aquatic nuisance.org/wordpress/wp-content/uploads/2009/01/ballast_water_issue _paper. pdf).

Pollutech. (1992). *A Review and Evaluation of Ballast Water Management and Treatment Options to Reduce the Potential for the Introduction of Non-native Species to the Great Lakes*. Pollutech Environmental, Ltd., Sarnia, Ontario, prepared for the Canadian Coast Guard, Ship Safety Branch, Ottawa, Canada.

Porra, R. J. (1991). Recent advances and re-assessments in chlorophyll extraction and assay procedures for terrestrial, aquatic, and marine organisms, including recalcitrant algae, 31-57. In: *Chlorophylls*, by H. Scheer (ed.). CRC Press, Boca Raton, FL.

Porter, J., Diaper, J., Edwards, C. & Pickup, R. (1995). Direct measurements of natural planktonic bacterial community viability by flow cytometry. *Applied and Environmental Microbiology*, *61*, 2783-2786.

Posch, T., Pernthaler, J., Alfreider, A. & Psenner, R. (1997). Cell-specific respiratory activity of aquatic bacteria studied with the tetrazolium reduction method, cyto-clear slides, and image analysis. *Applied and Environmental Microbiology*, *63*, 867-873.

Reavie, E. D., Cangelosi, A. A. & Allinger, L. E. (2010). Assessing ballast water treatments: Evaluation of viability methods for ambient freshwater microplankton assemblages. *Journal of Great Lakes Research*, *36*, 540-547.

Reeves, E. (1998). *Protection of the Great Lakes from Infection by Exotic Organisms in Ballast Water*. Report for Ninth US Coast Guard District, Cleveland OH.

Reeves, E. (1999). *Exotic Policy: An IJC White Paper On Policies for the Prevention of the Invasion of the Great Lakes by Exotic Organisms*. Documents on Water Law, College of Law, University of Nebraska - Lincoln. 132.

Rigby, G. & Taylor, A. (2001b). Ballast water management and treatment options. *Transactions of the Institute of Marine Engineers*, *113*, 79-99.

Rigby, G. & Taylor, A. H. (2001a). *Ballast Water Treatment to Minimize the Risks of Introducing Nonindigenous Marine Organisms into Australian Ports*. Ballast Water Research Series Report No. 13, Australian Government Department of Agriculture, Fisheries, and Forestry, Canberra, Australia, 93.

Romano, G., Ianora, A., Santella, L. & Miralto, A. (1996). Respiratory metabolism during embryonic subitaneous and diapause development in *Pontellamediterranea* (Crustacea, Copepoda). *Journal of Comparative Physiology*, *166*, 157-163.

Ropkins, K. & Beck, A. J. (2000). Evaluation of worldwide approaches to the use of HACCP to control food safety. *Trends in Food Science & Technology*, *11*(1), 10-21.

Rosacker, L. L. & Kieft, T. L. (1990). Biomass and adenylate energy charge of a grassland soil during drying. *Soil Biology and Biochemistry*, *22*, 1121-1127.

Ruiz, G. M., Smith, G. E. & Sytsma, M. (2006). *Workshop Report on Testing of Ballast Water Treatment Systems: General Guidelines and Step-wise Strategy Toward Shipboard Testing*. Aquatic Bioinvasion Research and Policy Institute, a joint program of the Smithsonian Environmental Research Center (SERC) and Portland State University (PSU). Submitted to the Pacific States Marine Fisheries Commission, Alaska Department of Fish and Game, and Prince William Sound Regional Citizens Advisory Council, Anchorage, AK.

Sandrin, T. R., Herman, D. C. & Maier, R. M. (2009). Physiological methods, 191-223. In: *Environmental Microbiology*, by R.M. Maier, I.L. Pepper, and C.P. Gerba (eds.). Elsevier, New York, NY.

Sarmento, H. & Descy, J.-P. (2008). Use of marker pigments and functional groups for assessing the status of phytoplankton assemblages in lakes. *Journal of Applied Phycology*, *20*, 1001-1011.

Schlüter, L., Lauridsen, T. L., Krogh, G. & Jørgense, T. (2006). Identification and quantification of phytoplankton groups in lakes using new pigment ratios - a comparison between pigment analysis by HPLC and microscopy. *Freshwater Biology*, *51*, 1474-1485.

Schmid, H., Bauer, F. & Stich, H. B. (1998). Determination of algal biomass with HPLC pigment analysis from lakes of different trophic state in comparison to microscopically measured biomass. *Journal of Plankton Research*, *20*, 1651-1661.

Shaw, S. & Kurunthachalam, K. (2009). Polybrominateddiphenyl ethers in marine ecosystems of the American continents: Foresight from current knowledge. *Reviews on Environmental Health*, *24*, 157-229.

Songster-Alpin, M. S. & Klotz, R. L. (1995). A comparison of electron transport system activity in stream and beaver pond sediments. *Canadian Journal of Fisheries and Aquatic Sciences*, *52*, 1318-1326.

Sperber, W. H. & Stier, R. F. (2009). *Happy 50th Birthday to HACCP: Retrospective and Prospective*. foodsafetymagazine.com (December 2009).

Steinberg, M. K., First, M. R., Lemieux, E. J., Drake, L. A., Nelson, B. N., Kulis, D., Anderson, D., Welschmeyer, N. & Herring, P. (accepted with revisions) Comparison of techniques used to count single-celled viable phytoplankton. *Journal of Applied Phycology*.

Steinberg, M. K., Riley, S. C., Robbins, S. H., Nelson, B. N., Lemieux, E. J. & Drake, L. A. (2010). Multi-site validation of a method to determine the viability of organisms ≥ 10 μm and < 50 μm (nominally protists) in ships' ballast water using two vital, fluorescent stains. *NRL Letter Report* 6130/1016, Washington, DC.

Stevens, T. G., Frederick, R. M., Everett, R. A., Hurley, J. T., Hunt, C. D. & Tanis, D. C. (2004). Performance verification of ballast water treatment technologies by USEPA/NSF Environmental Technology Verification Program. In: *Proceedings of the 2nd International Ballast Water Treatment R&D Symposium*, by J.T. Matheickal and S. Raaymakers (eds.). IMO, London, UK.

Suban, V., Vidmar, P. & Perkovic, M. (2010). Ballast Water Replacement with Fresh Water – Why Not? Proceedings from IMO and World Maritime University Research and Development Forum: Emerging Ballast Water Management Systems, January 26-29, Malmo, Sweden.

Sugita, H., Mitsaya, T., Amanuma, K., Hayashi, K., Asai, T., Maruyama, C. & Degechi, Y. (1992). Ultraviolet susceptibility of three marine fish pathogens. *Bulletin of the College of Agriculture and Veterinary Medicine*, *49*, 117-121.

Suttle, C. A. (2007). Marine viruses - major players in the global ecosystem. *Nature Reviews – Microbiology*, *5*, 801-812.

Suttle, C. A., Chan, A. M. & Cottrell, M. T. (1991). Use of ultrafiltration to isolate viruses from seawater which are pathogens of marine phytoplankton. *Applied and Environmental Microbiology*, *57*, 721-726.

Takano, C. T., Folsome, C. E. & Karl, D. M. (1983). ATP as a biomass indicator for closed ecosystems. *BioSystems*, *44*, 739-796.

Taylor, L. R. (1961). Aggregation, variance and the mean. *Nature*, *189*, 732-735.

Turner, R. E. (1978). Variability of the reverse-flow concentration technique of measuring plankton respiration. *Estuaries*, *1*, 65-68.

United States Coast Guard. (1999). Implementation of the National Invasive Species Act of 1996 (NISA). Interim rule. U.S. Federal Register, *64*(94), 26672-26690 (May 17, 1999).

United States Coast Guard. (2001a). Implementation of the National Invasive Species Act of 1996 (NISA). Final rule. U.S. Federal Register, *66*(225), 58381-58393 (November 21, 2001).

United States Coast Guard. (2001b). Report to Congress on the Voluntary National Guidelines for Ballast Water Management. Washington D.C.

United States Coast Guard. (2002). Standards for Living Organisms in Ship's Ballast Water Discharged in U.S. Waters: A Proposed Rule by the Transportation Department and the Coast Guard. 67 FR 9632, March 4, 2002.

United States Coast Guard. (2004). Mandatory Ballast Water Management Program for U.S. Waters. U.S. Federal Register, *69*(144), 44952-44961 (July 28, 2004).

United States Coast Guard. (2008). *Ballast Water Discharge Standard - Draft Programmatic Environmental Impact Statement*. U.S. Coast Guard, Washington, DC, 218.

United States Congress. (1990). Nonindigenous Aquatic Nuisance Prevention and Control Act of 1990. U.S. Congress, Public Law 101-6461, enacted on November 29, 1990.

United States Congress. (1996). National Invasive Species Act of 1996. U.S. Congress, Public Law 104-332, enacted on October 26, 1996.

United States Environmental Protection Agency (U.S. EPA). (1993). *Methods for the Determination of Inorganic Substances in Environmental Samples*. Report 600/R-93-100. Office of Research and Development, U.S. EPA, Washington, DC.

United States Environmental Protection Agency (U.S. EPA). (1997). *Method 445.0: In vitro Determination of Chlorophyll* a *and Pheophytin*a *in Marine and Freshwater Algae by Fluorescence*. U.S. EPA, Cincinnati, OH.

United States Environmental Protection Agency (U.S. EPA). (1999). *Criteria for Evaluation of Proposed Protozoan Detection Methods*. Report #EPA 815-K-99-02. Office of Research and Development, U.S. EPA, Cincinnati, OH. Available at: http://www.epa.gov/microbes/critprot.pdf.

United States Environmental Protection Agency (U.S. EPA). (2000). *Deposition of Air Pollutants to the Great Waters.* Third Report to Congress. U.S. EPA, Washington, DC.

United States Environmental Protection Agency (U.S. EPA). (2001). Aquatic Nuisance Species In Ballast Water Discharges: Issues and Options. Draft Report for Public Comment – September 10, 2001. Office of Water, Office of Wetlands, Oceans and Watersheds, and Office of Wastewater Management, U.S. EPA, Washington, D.C.

United States Environmental Protection Agency (U.S. EPA). (2002a). *Methods for Measuring the Acute Toxicity of Effluents to Freshwater and Marine Organisms*, 5th Edition. Report # EPA 821-R-02-012. Office of Water, U.S. EPA, Washington, DC.

United States Environmental Protection Agency (U.S. EPA). (2002b). *Short-Term Methods for Estimating the Chronic Toxicity of Effluents and Receiving Waters to Freshwater Organisms*, 4th Edition. Report # EPA 821-R-02-013. Office of Water, U.S. EPA, Washington, DC.

United States Environmental Protection Agency (U.S. EPA). (2002c). *Short-Term Methods for Estimating the Chronic Toxicity of Effluents and Receiving Waters to Marine and Estuarine Organisms*, 3rd edition. Report # EPA 821-R-02-014. Office of Water, U.S. EPA, Washington, DC.

United States Environmental Protection Agency (U.S. EPA). (2003). *Ambient Water Quality Criteria for Dissolved Oxygen, Water Clarity and Chlorophyll a for the Chesapeake Bay and Its Tidal Tributaries*. Report EPA 903-R-03-002. Office of Water, U.S. EPA, Washington, DC.

United States Environmental Protection Agency (U.S. EPA). (July, 2010). *Environmental Technology Verification Program (ETV) Draft Generic Protocol for the Verification of Ballast Water Treatment Technologies,* Version 5.1.U.S. EPA ETV in cooperation with the U.S. Coast Guard Environmental Standards Division (CG-5224) and the U.S. Naval Research Laboratory. National Sanitation Foundation International, Ann Arbor, MI, 62 pp.+ appendices.

Unnevehr, L. J. & Jensen, H. H. (1996). HACCP as a Regulatory Innovation to Improve Food Safety in the Meat Industry. *American Journal of Agricultural Economics*, *78*(3), 764-769.

Vaquer-Sunyer, R. & Duarte, C. M. (2008). Thresholds of hypoxia for marine biodiversity. *Proceedings of the National Academy of Sciences of the United States of America*, *105*, 15452–15457.

Veldhuis, M. J. W., Kraaij, G. W. & Timmermans, K. R. (2001). Cell death in phytoplankton: correlation between changes in membrane permeability, photosynthetic activity, pigmentation and growth. *European Journal of Phycology*, *36*, 1-13.

Vézie, C., Rapala, J., Vaitomaa, J., Seitsonen, J. & Sivonen, K. (2007). Effect of nitrogen and phosphorus on growth of toxic and nontoxic *Microcystis* strains and on intracellular microcystin concentrations. *Microbial Ecology*, *43*, 443-454.

Victoria Environmental and Natural Resources Committee (ENRC). (1997). *Report on Ballast Water and Hull Fouling in Victoria*. Parliament of Victoria, ENRC, Victorian Government Printer, Melbourne, Australia.

Vollenweider, R. A. (ed.). (1974). *A Manual on Methods for Measuring Primary Production in Aquatic Environments*, 2nd edition. International Biological Program Handbook No. 12, Blackwell Scientific, Oxford, U.K.

Wang, X. C., Qiu, F. G. & Jin, P. K. (2006). Safety of treated water for re-use purposes—comparison of filtration and disinfection processes. *Water Science & Technology*, *53*(9), 213-220.

Wetzel, R. G. & Likens, G. E. (2000). *Limnological Analyses*, 3rd edition. Springer-Verlag, New York, NY.

Wetzel, R. G. (2001). *Limnology*, 3rd edition. Academic Press, New York, NY.

Williams, S. C., Hong, Y., Danavall, D. C. A., Howard-Jones, M. H., Gibson, D., Frischer, M. E. & Verity, P. G. (1998). Distinguishing between living and nonliving bacteria: Evaluation of the vital stain propidium iodide and its combined use with molecular probes in aquatic samples. *Journal of Microbiological Methods*, *32*, 225-236.

World Health Organization (WHO). (2008). *Guidelines for Drinking-water Quality. Third Edition. Incorporating the First and Second Addenda. Volume 1: Recommendations.* World Health Organization, Geneva.

End Notes

[1] Assessing the Relationship Between Propagule Pressure and Invasion Risk in Ballast Water, 6/2/2011, http://www.nap.edu/catalog.php?record_id=13184)

[2] Protists refers to various one-celled organisms classified in the kingdom Protista, and which includes protozoans, eukaryotic algae, and slime molds.

[3] Dr. Cohen, who read these studies, objected to this sentence as being untrue and misleading and felt there had not been adequate opportunity for Panel discussion of the issue.

[4] EPA and the US Coast Guard have provided data they currently have to the panel. Where feasible, the panel is encouraged to find additional data if they have appropriate avenues to obtain those data.

[5] February 2011 Memorandum of Understanding Between the U.S. Environmental Protection Agency, Office of Enforcement and Compliance Assurance and the U.S. Coast Guard, Office of Marine Safety, Security and Stewardship for Collaboration on Compliance Assistance, Compliance Monitoring, and Enforcement of Vessel General Permit Requirements on Vessels (available at http://epa.gov/compliance/resources/agreements/cwa/mou-coastguard-vesselpermitrequirements.pdf)

[6] The remaining 5 management approaches involved neither shipboard treatment nor reception facilities.

[7] A protozoan pathogen with an active form measuring approximately 3 x 9 x 15 μm and an ellipsoid cyst averaging 10-14 μm long.

[8] A protozoan pathogen with round cysts 4-6 μm in diameter.

[9] Sequential combinations of some disinfectants produce reductions even greater than the sum of the disinfectants' reductions when examined separately (LeChevallier and Au 2004).

[10] EPA's current Vessel General Permit requires vessels on nearshore Pacific Coast voyages to conduct ballast water exchanges only if they cross international boundaries or cross from one Captain of the Port Zone to another (VGP §2.2.3.6). Similarly the U.S. Coast Guard's proposed discharge standards would not apply to vessels operating within a Captain of the Port Zone (USCG 2008c).

[11] Dr. Cohen, who read these studies, objected to this sentence as being untrue and misleading and felt that there had not been adequate opportunity for Panel discussion of the issue.

[12] Unless stated otherwise, the cost estimates cited in this appendix were converted from foreign currencies in the original publications into US dollars at the daily average interbank transfer rates reported at http://www.oanda.com/currency/historical-rates on the date of publication or presentation, or on the first day of the month where only the month of publication was given, and adjusted for inflation from the date of original publication to June 1, 2010 using the calculator at http://inflationdata.com/inflation/Inflation_Calculators/InflationCalculator. asp, which is based on the U.S. Bureau of Labor Statistics' Consumer Price Index for all Urban Consumers (CPI-U).

[13] Glosten (2002) designed the pumps and pipes to be large enough to enable ships to deballast completely at berth during a typical cargo loading period. Brown & Caldwell (2008) found, based on dynamic head vs. flow curves, that Great Lakes bulkers would not need larger ballast pumps—that is, with their existing pumps the ships could fully deballast while at berth during the time it takes to load cargo.

[14] Consistent with the study's aim of quantifying "the capital cost required to provide the maximum capability in a ballast transfer system, to represent a maximum capital investment" for each vessel category (Glosten 2002).

[15] This ability was included to accommodate the possibility of loading "clean" ballast, an approach that is not considered to be onshore treatment in this report.

INDEX

A

B

C

D

E

F

G

H

I

J

K

L

M

Q

R

S

T